DRIVING
WITHOUT GAS

JOHN WARE LINCOLN

GARDEN WAY 🌱 PUBLISHING
Charlotte, Vermont 05445

Printed in the United States

Portions of this book were originally published in 1976
under the title *Methanol and Other Ways Around the Gas
Pump*

Library of Congress Cataloging in Publication Data

Lincoln, John Ware.
  Driving without gas.

  "Portions of this book were originally published
in 1976 under the title Methanol and other ways
around the gas pump."
  Bibliography: p.
  Includes index.
  1. Alcohol as fuel. 2. Methanol. 3. Gas-
producers. 4. Synthetic fuels. I. Title.
TP358.L49   629.2'53   80-20655
ISBN 0-88266-172-8 (pbk.)

Contents

Preface

Energy is emerging as the number one domestic issue of the 1980s. Almost everywhere motorists are beginning to think about alternatives to gasoline. Words like Gasohol, ethanol and methanol are becoming more and more familiar. These and other alternate fuels — and innovative cars — are getting more and more attention. But work on these products, work on ways to *drive without gas*, has been going on for years.

The crisis in petroleum supplies in 1973 alerted the American public to the precarious relationship between world oil production and our prodigal driving habits. It also touched off a few thoughtful investigations of alternatives to oil-based motor fuels.

A group of scientists at Massachusetts Institute of Technology's Lincoln Laboratories, using their own family automobiles, rediscovered the merits of alcohol fuels in a series of tests that demonstrated their feasibility in conserving gasoline, improving performance, and reducing noxious emissions. I say *rediscovered* because Europeans, following World War I, took inventory of their meager oil supplies and used strong legal measures to extend them and to further the development of alcohol fuels, gasogens, and other alternatives. Conservation was urged by the enormous taxes on large cars. This was part of a planned defense against economic and military disaster.

Although this book was prompted by the activity of the Lincoln Laboratory pioneers, it draws heavily on historical material from my library, as well as the experiences of many sports car and steam car enthusiasts, hobbyists and professionals. Up-to-date information on the latest automotive developments has been provided by many generous individuals, including those mentioned here.

Some of the material dates from 1942, when my files on gasoline substitutes began to grow. Another time of research and collection came after the 1973–74 crisis at the pumps. The emphasis then was on methanol, and my book *Methanol and Other Ways Around the Gas Pump* was the result.

Much of that material appears here. Since 1978, there has been a wave of popular enthusiasm for Gasohol, a blend of 90 percent unleaded gasoline and 10 percent ethanol (grain alcohol). The potential of ethanol as a *long-range* alternative to gasoline does not equal that of methanol. Hence, the reader may find that methanol appears oftener, as the specific kind of alcohol fuel, whereas ethanol would have been appropriate, too. The difference, as it affects the performance of an engine, is slight, but methanol is far more difficult to produce.

Friends too numerous to mention have sent me clippings, tested my credibility and made suggestions. Charles E. MacArthur, Scott Sklar and William A. Stevenson gave me support, moral and physical, and Richard F. Merritt, head of the Alcohol-Alternate Fuel Institute, maintained the flow of Gasohol news to me from Washington, D.C. Dr. Thomas B. Reed, whose enthusiasm and perspicacity first stimulated this book, gave much time to reading and revising parts of the draft. Thanks to them all, and to my patient wife, Clarinda, the papers that have obstructed half our house for years have been compiled between these covers.

J.W.L.

Introduction

Petroleum

"Since early exhaustion of the supply is foreseen, it is worth remembering that the pursuit of happiness was not unknown before 1859."

(Columbia Encyclopedia, 1935)

In 1859 Col. Drake struck oil at Oil Creek, Pennsylvania.

After he became erect and abandoned his forelegs for locomotion, man was still limited in his mobility by the modest length of his hind legs. He could sprint briefly, a few hundred yards, after a deer, or he might extend his radius to a dozen miles, by jogging or walking. Early Greek storytellers, with wish-fulfilling fantasy, invented the *centaur*, combining a horsey speed and range with a human torso and the capacity to enjoy travel.

This need for better personal transportation was achieved, after a gestation of several millennia, in the motor bike and the automobile. To imagine that man will now lay down his new love-objects on demand, only because the price of fuel is going up, or because his money is said to be gravitating to far-off countries, is a folly comparable to national prohibition of alcoholic beverages.

When lines of cars formed at the pumps in 1974 and again in 1979, the degrees of irritation, anger, and violence that were manifested by the otherwise civilized people of the United States, approached those appearing during the Vietnam War. Although fuel for transportation is but one quarter of our liquid fuel consumption, it is the most sensitive area. Few motorists, and few politicians had any constructive ideas about how to escape from the terrible dilemma. Few could remember the 1930s, when a farm-produced alcohol fuel movement spread to about 2,000 service stations. These stations blended alcohol with gasoline. Almost none knew that about four million automobiles, trucks, and farm tractors in the United States, Europe, Latin America and the Philippines, have run, with no major mechanical alterations, on fuels derived from non-petroleum sources. This all occurred before 1938.

Peacetime Use of Alcohol

In peacetime, the use of alcohol as a motor fuel was sometimes fostered by its low price in the United States. Abroad, governments encouraged its use through high import duties on oil, subsidies for producers of "power" alcohols, and other supports of alternative fuel supplies, in the likely event of military emergency. In the United States, interest in alcohol fuels was intensified by rising gasoline prices, by stagnation in the liquor industry and by glutted grain markets, when corn sold for a dime a bushel, and farmers burned their grain to heat their houses or trucked it to a distant distillery, to barter it for motor fuel and some fertilizer and cattle feed.

Another strong force for alcohol fuel appeared in the form of a serious ecology movement: professors at the University of Iowa, and other grain state agricultural institutions, initiated legislation conducive to alcohol-blending with gasoline. They did not know that gasoline exhaust, inhaled over a period of time, was hazardous, but they did know that the replacement of the horse by tractors removed an important fertilizer from the farmers' dwindling store. Spent mash from distilleries was to become the manure from the iron horse.

It was unlikely that the professors knew that alcohol, even in a one-to-seven mix with gasoline, would reduce harmful exhaust emissions to a point nearly satisfactory to lawmakers 40 years later. This farm-grown fuel was ethyl alcohol (ethanol, or grain alcohol).

The reason for my emphasis on alcohol fuels (methanol and ethanol, in particular) is that they may be produced from a wide range of raw materials, and they may be mixed with gasoline to produce Gasohol, so that there need be no sudden shift in distribution methods. Unfortunately, this compatibility with our present fuel supply system and our rolling stock is not available to hydrogen-, methane-, or electric-powered vehicles

Because it can be made from cellulose wastes from forests and cities, from coal and peat, methanol has become the chosen alternative motor fuel in Europe and Scandinavia, where projects sponsored jointly by governments and manufacturers, such as Mercedes, Volkswagen and Volvo, have shown the feasibility of the fuel. Countries whose rich crops contain high sugar and starch fractions and do not compete with the food supply have chosen ethanol. Brazil, for example, plans to operate its transportation system almost entirely on ethanol by 1985.

In the United States, where there was a substantial underutilized capacity for ethanol production (now being absorbed rapidly for fuel instead of beverage production), Gasohol, a blend of 10 percent, high-proof ethanol with 90 percent unleaded gasoline, is the leading alternative fuel for the *near future*. Despite its controversial economic status—it is on the borderline of energy balance—its sales have blossomed fast; it is an excellent nontoxic octane booster; and it reduces, by a little, dependence on foreign imports. The proportion of alcohol may be increased, as new facilities come on stream.

Doomsday predictions have a way of softening with time, but it is quite clear that we are running out of oil and natural gas in the United States. Anxious readers may be consoled that motor fuels used in 1910, and now more desirable than ever, are being re-examined and found more than adequate. Volkswagen engineers have designed and tested engines for use with pure alcohol fuel. The reports demonstrate performance and economy superior to gasoline, without need for emission control gadgetry. Although there is no mention of the strategic value of homemade fuel in case Europe and the Western hemisphere are cut off from Middle Eastern oil, this is an important consideration.

Other Systems

While methanol, and in certain circumstances, ethanol, are practical potential extenders of the gasoline supply, there are other strategies and systems. Some are well known, like the electric vehicle, and others are only dimly remembered from the past. The *portable gas generator*, seen most frequently overseas, was one of the latter. It appeared all over Europe, on taxis, trucks, and private cars, to consume enormous quantities of wood chips, charcoal, coke, lignite, and plain coal. This miniature portable gas works, with a bit of contemporary technology, might be reduced in size and improved in convenience and efficiency to become the omnivorous goat of autodom, in case of an emergency. This unsophisticated creature has had little attention since 1946.

The layman's confusion about the energy problem is excusable. The magnificent, nostalgic steam car was revived in the early 1970s with much fanfare, because it was discovered to be nonpolluting. However, when it proved to be an oil hog, it was quickly reinterred.

Various Problems

Many other unorthodox automotive engines, and a few alternate fuels, such as methane and hydrogen, have been exposed to an eager public, but the engines can't be produced reasonably, and the fuels can't be burned or shipped, with economy and safety, and without disruption of a huge segment of the personal and commercial lives of wheeling Americans, particularly those in the one-sixth part of United States population connected with the traditional, tightly controlled systems of distribution in automotive, fuel, and associated service industries. One exception — there may be others — is alcohol, a renewable fuel, blended with gasoline.

Although the change from petroleum to other fuels made from renewable sources rather than from fossil material is the dominant theme of this book, it must be obvious to temperate observers of the energy problem that no single fuel substitution, no technical breakthrough, no specific rule of conservation, nor any superimposed economic theory, will enable the American people to continue their once-carefree life with a full tank of gasoline at a dollar per gallon. Only a compromise can be achieved, and then only by the use of every appropriate technical, legal and economic strategy available to us.

The doom of the private automobile is not sealed. I am heartened to see how many farmers are making their own fuel. I rejoice to read, frequently, that Jack Smith, of Jones' Garage, has driven his old sedan 80 miles on a gallon of Gasohol with a carburetor he made himself. I find good news for New England and southern motorists in a report from a Scandinavian car manufacturer experimenting with turpentine, a fuel that may become an important product of softwood forests.

CHAPTER 1

Gasoline: Going, Going . . .?

"There is no real shortage of gasoline. I don't believe it. They tell us this to justify raising the price. There's enough oil in the ground to last long after I'm dead."

If you are more than 70 years old, there may be a little truth in that oft-repeated assertion, but if you are 20, your span of unrestricted private-vehicle driving, with a tankful of petroleum, is sure to be limited. You will adjust to scarcity, and to prices we now consider prohibitive (more than the $3 per gallon some Europeans pay). As time passes, the fuel in your tank will contain more and more alcohol, made from renewable crops, wastes from cities and forests (biomass) or from coal. Before you reach middle age, cars will be made with small, efficient, high-compression engines and fuel injection, expressly designed to run on pure alcohol fuels. Several of these are operating in the United States, Sweden, Germany and Brazil at present.

Motorists' Misconceptions

The complacency of some American motorists about the reality of an oil shortage is based on three general misconceptions. The first is the Breakthrough Theory: our wonderful technologists somehow will devise ways to make cheap fuel — from shale, from sea water or from nuclear reactors. Hydrogen is mentioned as the new miracle fuel. Some of these possibilities are examined in this book, but the outlook is hardly promising. The second misconception is that the same clever scientists who may achieve a breakthrough are afflicted with a blind spot. They have miscalculated the extent of petroleum reserves and are crying wolf. The third misconception is that the petroleum industry is withholding supplies of gasoline to raise the price artificially. Apathy and confusion about the

1

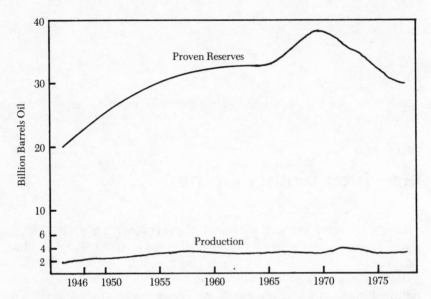

Figure 1-1. If proven U.S. reserves continue decline begun in 1971, new energy alternatives must be sought and developed. SOURCES: *American Petroleum Institute, Canadian Petroleum Institute.*

conflicting pressures of industrial energy consumers and the forces of environmental protection do not contribute to an understanding of and sympathy for the oil shortage.

Forecasting Fuel Supplies

It is important, in seeking a reliable and useful estimate of future oil supplies, to understand that changes in production of domestic oil occur very slowly, unlike imported supplies, subject to abrupt political cutoffs. Even more important is the relationship of oil production to the estimates of proven reserves. There is a gap of 8 to 12 years between a new find and actual production. Figure 1-1 shows the relationship of proven United States reserves to actual production over a 31-year period. (In 1946, new finds were increasing the reserves by 1¼ barrels of oil per barrel produced from wells. The ratio now is a half-barrel found to one being pumped.) When projected to the year 2000, the chart lines suggest increasing dependence on imported foreign oil. But, the availability of these imports, even at prices double the present ones, is clouded by uncertainties.

More Fuel Demand

The downward slope of the reserve line and the nearly level trend of production would be serious enough even if transportation and industry needs were to remain at present levels. To keep production adequate to this need requires a discovery rate 50 percent greater than that for the last 10 years. But the United States vehicle fleet is increasing at an annual rate of 4.2 percent. Truck registration alone increased at 6.2 percent over the decade 1967–1976. Trucks averaged 8.5 miles per gallon in 1976, and in another decade, with 62 percent more of them, it is difficult to visualize any immediate solution to the expected shortfall other than a huge increase in foreign imports.

Only a crisis situation seems to stimulate the American public to action, and political leaders are prone to wait for an energy crisis rather than anticipate. The technological community, by contrast, recognizes the folly of waiting. The retiring president of the American Association for the Advancement of Science (AAAS), Roger Revelle, compared this faith in the efficacy of crash programs to a belief that producing a baby in one month can be accomplished by putting nine women on the job.

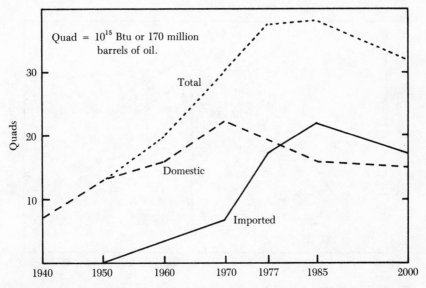

Figure 1-2. Annual U.S. liquid petroleum supplies, from both foreign and domestic sources, are expected to decline, according to this forecast by Earl T. Hayes, former chief scientist, U.S. Bureau of Mines. (Science, vol. 203, January 19, 1979.)

Reserves Are Vulnerable

While Figure 1-1 shows reserves greater than production, there is no cause for complacency. The figures for reserves include estimates of petroleum deposits in shale, deep wells, and submarine fields. Drawing on these deposits will escalate the price because the oil will be difficult to extract and refine.

The vulnerability of the reserves is shown by the statistics for the most recent year reported, 1978. At the beginning of that year, proven reserves were 29.5 billion barrels. New finds added 1.3 billion barrels, but withdrawals by the producing industry were 3 billion barrels. Hence, the net loss to reserves was 1.7 billion barrels. It is not difficult to foresee the results of this kind of depletion in a couple of decades. Figure 1-3 shows the estimates of production of petroleum liquids over such a period. As domestic production is forced down by increasing difficulties of extraction, the pressure for greater imports must continue to increase. Imports pose an economic problem — the deterioration of our trade balance — as well as a strategic impasse: the reserves in the OPEC fields are finite; most of the

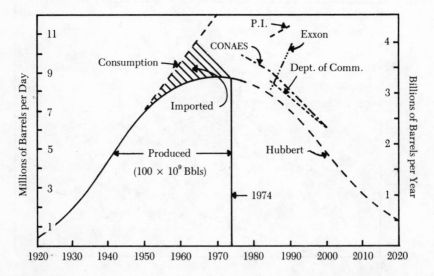

Figure 1-3. American petroleum consumption has exceeded production since the early 1950s, and the deficit has been filled by imported oil. Various authorities, including the Committee on Nuclear and Alternative Energy Systems (CONAES), the Department of Commerce and M.K. Hubbert, project a continued production decline in the decades ahead. Exxon and Project Interdependence (P.I.) are more optimistic.

oil has already been found, and world production in the last decade has taken half of the oil ever produced in the world's history. It would be most imprudent to assume that OPEC will supply the United States with unlimited oil at any price.

The Solution

The solution, unless it be a reversion to the animal and human energy sources of the nineteenth century, *must be found in the gradual substitution of liquid fuels made from renewable raw materials, or abundant fossil solid fuels, coal and shale.*

The demand for liquid fuels during the transition period will be modified by economic forces. Growth rates for gasoline consumption, the subject of much disagreement because of flexibilities in demand, have been estimated at 2 to 3 percent per year. At these rates, by 1990, daily consumption of 9 million and 11 million barrels could be expected. But some forecasters recognize several factors that lead them to reduce estimates of daily consumption. Among them:

- U.S. population growth is slowing. Five of the most highly industrialized countries in Europe had dropped below the zero-growth point by 1976. By 2000, the United States rate may be under 1 percent per year.

- Price increases and voluntary conservation will reduce consumption of fuel.

- Automobile ownership, now at 2 to 3 cars per family in many cases, will be reduced due to excessive costs. The self-regulatory effect of traffic jams ("fatigue-rationing") will be seen.

- Fuel efficiency of engines will increase from 20 to 50 percent by 2000. Lighter cars and better materials as well as improved engine design will be instrumental.

- Miles traveled each year by the average driver will go down from 9,800 now to 9,300 per year.

- Electric vehicles are expected to reduce gasoline needs.

All of these factors, *if applied*, could reduce the annual growth rate of gasoline consumption to about 1 percent. But forecasting is not an exact science. In fact, it is more like clairvoyance: part knowledge, part art. For example, in the face of predictions of steady growth, gasoline consumption actually dropped in 1979 to about 7 percent below 1978 levels, according to John C. Sawhill, deputy secretary of the Department of Energy.

If uncertain fuel supplies and ever-increasing prices darken the road ahead for motorists, the history of the internal combustion engine offers some consolation and reassurance. It is a 100-year record of adaptation to new ideas, to new markets; it is the story of improved modes of propulsion, of ingenuity in meeting constraints on sources and supplies. And there is no reason to think alternate fuels and innovative cars can't be an even larger part of the story in the years ahead.

CHAPTER 2

Gasohol and the Revival
of Grain Alcohol (Ethanol)

Gasohol is an idea that has been smoldering for years in the nation's corn-belt, and now it is sweeping out to all corners of the country — a bit like an out-of-control brush fire. More than 2,000 service stations sell Gasohol. In the farming regions of the midwest, automobile bumper stickers boost this newly named fuel that is derived, in part, from corn: SUPPORT AMERICAN HOME-GROWN FUEL, says one; PUT AMERICA BACK IN THE DRIVER'S SEAT: USE GASOHOL and MY CORNFIELD IS YOUR OILFIELD are others.

For many, Gasohol is nothing more than an old idea whose time has come. The term *Gasohol* is defined as a mix of 90 percent unleaded gasoline and 10 percent (200 proof) ethyl alcohol (ethanol). Corn is the most common fermentation feedstock for the manufacture of ethanol used in Gasohol. During World War II, alcohol fuel mixes were used in Europe to help conserve petroleum. Even earlier, in the 1930s, Nebraskans pressed for legislation encouraging alcohol fuel production. At the urging of Gov. Charles Bryan, the state legislature passed a bill that allowed farmers tax rebates on alcohol blends. Iowa and other grain-raising states passed similar legislation in the 1930s and 1940s.

When "cheap" petroleum fuel became widely available after World War II, interest in alcohol fuels faded. But now, hundreds of thousands of gallons of grain alcohol are produced for fuel every day. One of the nation's largest producers, the Archer Daniels Midland Corporation of Decatur, Illinois, ships alcohol to more than 40 states. According to the Department of Energy, ethanol production for fuel use is expected to exceed 500 million gallons by 1985. In the meantime, Gasohol, the best known alcohol-gasoline fuel, is gaining ever-wider acceptance as more and more motorists become frustrated and fed up with high-priced, imported petroleum products (Figure 2-1).

7

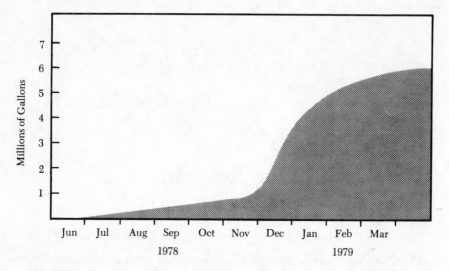

Figure 2-1. Gasohol sales increased dramatically in the state of Iowa, as shown in this graph adapted from data provided by the Iowa Development Commission. One key factor was the implementation in January 1979 of a federal tax exemption of 4 cents a gallon.

HOW DOES GASOHOL PERFORM?

Even the big three automakers are taking a close (but not overly enthusiastic) look at Gasohol. General Motors, Ford and Chrysler say the use of Gasohol containing up to 10 percent ethanol does not invalidate warranties on their vehicles. Even though some manufacturers are developing and producing cars to operate on pure, 100 percent ethanol, most suggest that motorists contact their dealers before putting straight alcohol in their tanks. The implication is that cars so powered would not be fully warranted.

Fuel Economy

Mileage test results for cars using Gasohol are not conclusive. For example, the Illinois Bell telephone company compared 15 vehicles using Gasohol with regular gasoline. The results: "The vehicles using Gasohol averaged 4.43 percent better mileage. Our monthly maintenance review of the vehicles also indicated some savings in engine maintenance with the cleaner-burning Gasohol," said the company's manager for automotive services. "We also found that Gasohol virtually eliminated gas line freezing problems during the winter months."

Other studies have said cars powered by Gasohol attain 3 to 5 percent better mileage. One well-known study, The Nebraska Two-Million-Mile Gasohol Road Test Program, reported that Gasohol-fueled cars got an average of 5 percent more miles per gallon than cars using unleaded gasoline. And the Land O'Lakes corporation of Fort Dodge, Iowa, reported that different motorists' driving habits affect mileage, but test results with fleets of up to 100 cars showed a 3 to 5 percent increase.

General Motors, though, which has experimented with ethanol seriously for several years, has said, "For the average vehicle, fuel economy will be lower with Gasohol than with gasoline because of Gasohol's reduced heating value . . . for most vehicles, the fuel economy *penalty* (emphasis added) for Gasohol will be in the range of 0 to 3 percent."

Some Detroit engineers contend that ethanol contains only two-thirds of the energy measured in Btu of the gasoline it replaces. What they may sometimes fail to mention is that the two-thirds represents 60 percent of

Motorists in most of the states can find a service station providing customers with Gasohol, a blend of unleaded gas and alcohol. (Iowa Development Commission)

the ethanol used in Gasohol, and the need for energy in the form of gasoline is reduced by that amount. Also, as pointed out by the *Conference Board*, an authoritative business journal, ethanol combusts more efficiently than gasoline. About a third of the weight of an ethanol molecule is oxygen, and that makes for better burning and peppier engines, especially in older cars.*

Vehicle Performance

Most alcohol fuel enthusiasts and experts agree on at least one point: Gasohol has an average octane rating that is three or four points higher than plain unleaded gasoline. This higher rating means that engine knock and pinging are eliminated in many cars. But Joseph Colucci, fuels and lubricants department chief for GM, contends that the higher octane of Gasohol will not improve fuel economy because most car engines are not designed for it. In the future, Colucci says, if Gasohol use becomes widespread, engine compression ratios might be increased to take advantage of the higher octane quality of Gasohol. More likely, the oil companies might utilize the ethanol as an octane-boosting agent. This would permit them to decrease the severity of their gasoline refining operations, and thus decrease refinery energy expenditures.

Actually, increasing compression ratios, by "shaving" cylinder heads, is nothing new. Just before World War II, I acquired a used English sports

---

*James Krohe, Jr., "Gasohol: It Has Turned Skeptics Into Believers. Well, Some of Them," *Conference Board*, October 1979.

WHAT IS OCTANE NUMBER?

Early in the development of the modern internal combustion engine, it became necessary for engineers to know the quality of liquid fuels and to design engines that could best use their energies. Many liquids and gases were tested in a one-cylinder laboratory engine that had a variable stroke, so that piston travel and hence compression ratio could be changed at will.

A liquid called *heptane* would cause knocking at any compression ratio, while *iso-octane*, another hydrocarbon, would not knock at all. When an unknown gasoline was to be tested, the engine's compression ratio (C.R.) was increased, while running on this fuel, until it began to knock. Then the engine, with the same C.R. setting, was run on heptane, and octane was added by a mixing valve until the knocking faded away. If this occurred when the mixture was 70 percent octane, the unknown fuel could be labeled 70 octane.

FIGURE 2-2. COMPARISON OF IMPORTANT FUEL PROPERTIES

Properties	Typical Gasoline	Ethanol	Gasohol
Oxygen content (Wt. %)	0	35	3.7
Chemically correct air-fuel ratio	14.5	9	13.9
Energy content (Btu/gal.)	114,000	75,000	110,000
Vapor pressure (lbs./in.^{2})	10.0	2.2	10.7
Octane quality*	87	98	90

*Average of research and motor octane numbers.
SOURCE: General Motors Research Laboratories.

car. Its engine had been "souped up" by shaving off the cylinder block to raise the compression ratio to more than 8:1, giving the engine more compression than most cars of that time. Although I used the highest octane premium gas available, the engine knocked at the slightest acceleration.

At a gathering of racing enthusiasts, an English owner of a similar car advised me to give it alcohol. Thereafter, with about 10 percent methanol, an alcohol fuel with an octane number estimated to be over 100,* I had no more knocking and ventured forth on an oval track to race. My car reached speeds of 85 mph regularly, and a rough computation indicated that the little engine, no larger than the one in a Model T, produced six times the original horsepower when in trim and when operating on the alcohol blend.

Even if the notion of high speed in an automobile is in disrepute, a small engine with a high compression ratio is much more efficient in weight, cost and fuel consumption than a large, sluggish engine. Without high-octane fuel, the modern car could not have evolved. The British, with an oppressive tax on cars with engines of large dimensions, are to be thanked for evolving the light, high-compression ratio, sports-car engine.

Tetraethyl Lead. Before the advent of Gasohol, tetraethyl lead was the standard additive for increasing octane number. It was such a good booster that gas of 50 octane could be raised to 60 by adding 1 cubic centimeter of lead per gallon; 2 cc would push it to 68, and 3 cc achieved a

*Pure methanol has an octane of 106.

maximum of 72. Most of today's refineries were designed to produce the largest possible amount of 60- to 70-octane gas. This was adequate, with lead additive, for the newer cars, up to the beginning of the war on air pollution.

The war on air pollution was, in part, a war against lead. Lead residues accumulate along roadsides, and they represent a serious health concern. There have been frightening studies in urban areas of Massachusetts of lead concentrations in childrens' teeth. Small amounts of body lead are related to learning disabilities and serious childhood diseases.

Catalytic converters help break down and reduce harmful exhaust emissions. But, unfortunately, *catalytic converters cannot tolerate lead*. And the refineries are still struggling to improve their processes to raise the natural octane number. They have tried the additive MMT (methyl-cyclopentadienyl manganese tricarbonyl), which also seems to have a bad effect on catalytic converters. It was phased out of production in September 1978.

Tetraethyl lead is to be banned as a fuel additive, on a lingering schedule that began in January 1978. Under these restrictions, there's insufficient lead to provide enough octane boost for the new high-performance cars. Therefore, the three largest oil retailers have pulled some old tricks out of the racing drivers' medicine kit to extend octane without lead. They've added up to (and in some cases more than) 15 percent benzine to certain premium unleaded fuels. Although benzine is an excellent octane booster, it is also a carcinogen. Amoco and Arco premium fuels are said to be fortified with 5 to 7 percent tertiary butanol — a fine, higher alcohol produced from petroleum.

Thus alcohol is the only practical fuel additive that gives a genuine octane boost without increasing pollution (it lowers it, in fact). Also, alcohol is not carcinogenic, not imported, and not harmful to catalytic converters.

Separation problems. When small amounts of water accumulate in a fuel tank containing an alcohol-gasoline blend, the two components tend to stratify as the temperature drops. The alcohol settles at the bottom of the tank. When it is pumped up for a cold start in the morning, the alcohol fails to vaporize in the carburetor, and it is necessary to heat the inlet manifold, by an external or internal electric heater.

Another option is a primer that provides a brief starting injection of propane gas, ether or volatile gasoline. Such a primer is expected to be a routine accessory on the future cars designed to burn 100 percent alcohol, but it should not be necessary for the transition period ahead, when blended fuels and unmodified engines must be coordinated for the North American climate.

Nebraska test. In the Nebraska Two-Million-Mile Gasohol Road Test Program there was little dissatisfaction with car performance. Drivers said they experienced no hard starting or stumbling of the engines. Furthermore, they said, the Gasohol-fueled cars appeared to have better pick-up. No vapor lock was reported, and overall driver satisfaction with car performance was high.

Exhaust Emissions

One of the finest things about burning alcohol in an automobile engine is that the exhaust emissions are generally far cleaner than from regular unleaded or leaded gasolines. Almost all automotive experts agree that Gasohol-powered cars emit less carbon monoxide — some studies say as much as 25 percent less. But, Gasohol may either increase or decrease hydrocarbon emissions and nitrogen oxides emissions, depending on how the engine is calibrated initially. The Nebraska test showed a total reduction of all emissions, but the total amount for Gasohol was only slightly less than that for unleaded gasoline. (See Figure 2-3.)

FIGURE 2-3. SUMMARY OF ERDA* EMISSIONS AND FUEL CONSUMPTION TESTS AT 75° F. (24° C.)

	GRAMS PER MILLION					
	CAR 53003		CAR 53000		2-CAR AVERAGE	
Component	**Gasohol**	*No-Lead*	**Gasohol**	*No-Lead*	**Gasohol**	*No-Lead*
CO	15.4	23.1	25.9	38.1	20.7	30.6
HC	1.8	2.0	2.8	2.7	2.3	2.3
NO_x	2.4	2.7	2.1	1.8	2.3	2.3
Subtotal	19.6	27.8	30.8	42.6	25.3	35.2
CO_2	751.9	768.9	658.7	653.4	705.3	711.1
Total	771.5	796.7	689.5	696.0	730.6	746.3
Fuel Consumption						
Urban Mi/Gal	10.4	10.4	11.5	11.7	10.9	11.0
Hwy Mi/Gal	15.1	15.5	16.3	16.7	15.6	16.1
Combined Mi/Gal	12.1	12.2	13.3	13.5	12.7	12.8

*Energy Research and Development Administration

Engine Parts and Other Considerations

GM and many others that have experimented with Gasohol have *not* found that using the new alcohol-gasoline blend causes any material or component failures in automobiles. While GM cautions that more experience with the fuel is needed before its long-range impact on cars and their parts can be determined, the big three automakers are not invalidating their car warranties.

From my experience with alcohol fuels, and from the reports of tests by others, it is safe to conclude that there are a few results that can be expected from using an alcohol blend in a car:

1. Ethyl alcohol or ethanol tends to dissolve gook and sometimes frees rust in fuel lines and in the fuel tank. The fuel system is cleaner after a tankful or two, but the fuel filters often must be either cleaned or replaced.

2. As shown in the Nebraska test, Gasohol is even safer than pure alcohol, which may damage certain parts. Gasohol and unleaded gasoline produced no difference in engine cylinder wear. There were no more carbon deposits or wear to spark plugs, valves or valve seats and no premature failure of any engine parts or fuel-line components.

ETHANOL: RUNNING CARS ON "STRAIGHT" ALCOHOL

In Brazil, where there are acres and acres of sugar cane, an ideal agricultural product for making alcohol, there is also a national commitment to develop and manufacture automobiles that operate on nothing else. By 1982, Brazil expects to produce or modify nearly 1.1 million cars to operate on 100 percent alcohol. Seven hundred test vehicles already operate on pure alcohol and many more are using alcohol blends. Ultimately, Brazil plans to manufacture only cars that operate on pure alcohol.

This is a major international program, involving American, Italian, German, Swedish, and Japanese automakers. Ultimately, the program will help to free the Latin American country from its dependence upon petroleum. It also will be a major boost to Brazil's agricultural industry.

Brazil is leading the world in the production of alcohol-fueled cars and Americans can operate their own cars with straight alcohol as well. In fact, ethyl alcohol, or ethanol, with a proof of only 160 (80 percent alcohol; 20 percent water) will work in simply modified cars. More and more car enthusiasts and experts are experimenting with ethanol — in cars, tractors and even power boats. To operate a combustion engine on ethanol, a few modifications are necessary.

Engine Modifications

It is possible to operate many cars on the road today with pure ethanol if they are equipped with properly adjusted carburetors, and if they are able to vaporize enough fuel to get the car started at temperatures below 50° F. Also, ethanol is more corrosive than gasoline to some fuel filters,

An industry leader in the field of alcohol-fueled cars,
Volkswagen has designed this engine to operate on ethanol.

lines and tanks, meaning that these items might have to be replaced or modified. But these problems are far from insurmountable. Carburetors can be modified; durable parts can be substituted for those subject to corrosion.

Carburetor and fuel pump. Ethanol of 160 proof or better can be used in most ordinary cars if the flow of fuel to the engine is increased. Fuel must be increased because the heat values for ethanol are lower than those for gasoline. Hence, the carburetor fuel jets must be enlarged, usually by about 40 percent. The fuel pump, now called upon to deliver twice as much volume as before, also must be augmented.

Starting device. Another desirable addition, and one that is a necessity if you live in the northern United States, is some form of device for starting in cold weather. This device can be either a block heater, or a propane gas cylinder, solenoid valve and a pressure-reducing valve. The propane system is designed to inject the flammable gas into the air cleaner when you turn on the car starter switch. While propane gas will remain in the vapor state until it reaches the cylinder, alcohol becomes more difficult to vaporize as the thermometer drops below 50° F.

Preheater. Even when vaporized in the carburetor, alcohol may condense on the cold walls of the intake manifold. Compared with gasoline, alcohol produces larger droplets and rapid coagulation that makes proper distribution of the fuel-air mixture to each cylinder more difficult. This is the reason for the gas starter, and for the addition of air preheating. The latter is accomplished by passing the incoming air over part of the exhaust system. Light sheet metal scoops are the usual solution, while flexible metal ducting, of a diameter somewhat greater than that of the carburetor inlet, may also be run against a section of exhaust pipe.

The best place for an air intake heater is usually along the short but hot section of the exhaust manifold. Most air cleaner intakes can be rotated, and an inspection of your exhaust system will indicate that a three-sided channel of sheet metal or a piece of flexible metallic duct, leading air *back*wards from the radiator, over the exhaust pipe, can be connected to the air cleaner horn (intake). These connections, and the fit of the scoop over the exhaust pipe, need not be perfect fits, up to the cleaner. The tuning of the carburetor, with and without the air cleaner, will be different.

Another way to modify an ethanol-powered engine to improve its performance is to raise the engine's compression ratio, and adjust the timing. Raising the compression ratio, while not essential, is one of the best changes to make to achieve improved performance and fuel economy. This job is best done by an experienced machinist or auto mechanic.

ECONOMIC AND SOCIAL ISSUES

The State of New York studied a fleet of cars powered by Gasohol and then issued an 18-page report. Included was what the authors of the report called a "provocative review of the whole Gasohol issue." The word "provocative" hinted at the tone of the review and its likely conclusions; the opening quote was a pure giveaway:

> "One of the great tragedies of life is the murder of a beautiful theory by a gang of brutal facts."
> BENJAMIN FRANKLIN

If you've read this far, you probably need no further coaching to guess what this "review" established about the notion of fueling cars with corn power. In short, the report by the state's agriculture department, concluded that Gasohol produced from *feed products* would be nothing less than a "disastrous food and farm policy."

But for every New York state agriculture department, there are other departments or reports with the opposite view — that making alcohol fuel from corn is a sensible, timely idea. At the same time, dozens, hundreds, perhaps thousands of scientists, researchers, experimenters, sociologists and even politicians are trying to determine just exactly what the "brutal facts" are about alcohol fuels. Opinions are sharply divided. Definitive statements are scarce or outdated. But there are a few persons willing to argue on both sides of the many economic, social and political issues that swirl about the words Gasohol and ethanol (ethyl alcohol).

Alcohol Production Costs

There are at least two ways to examine the costs associated with ethanol — from a production viewpoint, and from the viewpoint of the consumer. What does he pay for an ethanol blend such as Gasohol when he drives up to the gas pump? Many researchers and fuel experts agree that the energy consumed in the production of a gallon of ethanol is as great as the energy derived from that gallon. But, others ask, isn't that true of conventional fuels as well? And just exactly what is included in production costs?

Again, few agree on what should be included in a fair assessment of the costs of producing ethanol. But, however you look at the costs, no matter what you include in the equation, it's always true that whenever we convert from one form of energy to another—from corn to alcohol, or from coal to electricity—there is an energy loss. In this kind of conversion, ethanol doesn't do too badly. For example, compare corn-to-ethanol to another conversion we have tolerated for years—coal-to-electricity.

CONVERTING CORN TO ETHANOL

Input		Output	
To grow corn (2/5 Bu.)	41,000 Btu	1 gallon ethanol	84,000 Btu
To ferment and		Dried grain	
distill	131,000 Btu	(by-products)	50,000 Btu
Total	172,000 Btu		134,000 Btu
		Deficit:	38,000 Btu
		Efficiency:	78 percent

COAL TO ELECTRICITY

Input		Output	
1 lb. coal	10,000 Btu	Electric equal	3,413 Btu
*(Mining energy		Deficit:	6,587 Btu
not available)		Efficiency:	34 percent

*If the Btu losses in mining coal were included, it could be shown that generating electricity with coal is a totally useless operation. If the value of human lives lost could be included in mining costs, the situation would be far worse.

Other studies suggest that ethanol can be produced even more efficiently with new and improved technology or with other feedstocks besides corn. For example, one study conducted by Battelle Columbus Laboratories for the American Petroleum Institute, the oil industry's major organization, determined that "ethanol from sugar cane" is a "net energy producer." Methanol, if produced with processes now under consideration, could also be a net energy producer. (See Figure 2-4.) The Battelle study considered all energy costs associated with production—fuel to grow feedstocks, fuel to make fertilizer, and fuel to run the alcohol production process.

But Battelle's conclusions about sugar cane differ slightly from those of a research team at Louisiana State University's Coastal Ecology Laboratory. The LSU researchers determined that about 2.7 acres of sugar cane could produce alcohol worth about 18.4 million calories of energy. This, they calculated, would be about 900 calories less than the energy required

FIGURE 2-4. ALCOHOL FUELS ENERGY EFFICIENCY

	*Efficiency**
Ethanol from sugar cane	300 percent
Ethanol from corn (traditional technology)	45 percent
Ethanol from corn (energy-conserving technology)	83 percent
Ethanol from corn stover	65 percent
Methanol from wood (Purox process)	217 percent
Methanol from wood (Battelle process)	222 percent

*These efficiency percentages were determined by applying a standard efficiency formula, in this case Btu output divided by Btu input (the energy needed to produce the alcohol fuel). The percentages were calculated from base data furnished by the American Petroleum Institute.

to produce the alcohol. This appears to be a small energy loss, but other studies have produced even more unfavorable equations for ethanol.

One way to cut ethanol energy-production costs would be to use crop residues such as corn stalks as a fuel for the ethyl alcohol distillation. But Gasohol detractors are not at all pleased by this suggestion, either. They point out that massive use of agricultural wastes as a fuel would hasten soil depletion. And many are horrified at the thought of using anything like corn to run automobiles. Here's a typical analysis from an opponent of alcohol fuels from corn:

> United States gasoline consumption is 100 to 110 billion gallons yearly. A 10-billion-gallon Gasohol program would use 1 billion gallons of alcohol. To produce this amount of alcohol from corn would use over 50 percent of present corn reserves and over 280 percent of the corn which farmers hold under the federal loan program. This program level would only displace 0.8 percent of our current imported crude oil use. To displace 8.2 percent of our current crude oil imports would take 57 percent of all current corn acreage.*

Of course, an immediate transition to such a massive corn-alcohol program is unlikely, if not impossible. But there are many thoughtful people who question the wisdom of shifting the purpose of corn from food to fuel. One fact they may sometimes overlook is that the government in 1978 alone paid more than $1.6 billion to farmers to compensate them for *not* planting crops.

Report on the Testing of Alcohol/Gasoline Fuel Blends in Conventional Vehicle Fleets in New York State, Office of General Services, Division of Interagency Transportation Services, New York State, March 1980.

The Department of Energy predicts that the feedstocks for alcohol fuels in the 1980s will probably be wastes from agricultural distressed products and by-products. Cellulose materials not useful as food, and coal and peat could be processed in the future to produce ethanol and methanol, respectively.*

In the meantime, researchers are looking for ways to assure a bright future for ethanol by reducing the costs of production. For example, Michael Ladisch and Karen Dyck of the Laboratory for Renewable Resource Engineering at Purdue University announced a method of reducing the major energy requirement of distillation (50 to 80 percent of total manufacturing heat) of ethanol to the high proof needed for Gasohol. Instead of lengthy redistillation of the "beer," a mixture of water and alcohol vapors, the Purdue team passes these over cornstarch at a moderate 90 degrees centigrade. The water is absorbed by the starch, while the alcohol is rejected and condensed. The wet starch is then used, without drying, in the feedstock. The combustion energy of the product can be 10 times greater than the energy needed to dehydrate it.

The Cost to Motorists

To the surprise and bewilderment of some oil industry executives, more and more motorists are buying Gasohol — even though it often costs at least a few cents more than regular or unleaded gasoline. Why? Because Americans prefer a domestically produced fuel that they believe will help reduce foreign imports, will keep gasoline costs down, and will do the least damage to the land. Presumably, they see coal extraction as more damaging. These were among the findings of a Louis Harris poll of 7,010 Americans 18 years of age or older.**

In an earlier study in Iowa, consumers said they liked Gasohol because they believed it extended oil supplies, improved automobile performance, and helped the corn market. Only 21 out of 514 users of Gasohol said they encountered problems. "Rougher-running engines and plugged fuel filters, which are not unusual problems in initial tankfuls of Gasohol, were experienced by 10 of 21 users."†

---

*The Report of the Alcohol Fuels Policy Review, U.S. Department of Energy, Washington, D.C., June 1979. See Appendix B for a discussion of this report.

**Louis Harris and Associates Survey conducted between October 19 and November 21, 1979.

†"Gasohol Acceptance in Established Markets," Iowa Development Commission, April 19, 1979.

One reason Gasohol remains competitive is that the price of gasoline is increasing and Gasohol is getting help from Uncle Sam in the meantime. Ethanol costs nearly twice as much as unleaded gasoline at the wholesale level. But by the time the ethanol-gasoline blend reaches the gas pump, it costs only a few cents more than the traditional fuel. The government exempts the sale of Gasohol from the four-cent-a-gallon federal gasoline tax. And several states have tax reductions in effect for alcohol fuels.

There may be political, as well as economic reasons for motorists to fill up with Gasohol. It doesn't seem to be a coincidence that much of the support for Gasohol comes from the corn belt and other southern states where self-sufficiency and independence from foreign petroleum are on the minds of many. This sentiment was captured quickly by a new firm in Buena Vista, Georgia, marketing a portable alcohol-fuel distillation unit. The unit was called, in large letters painted across its side, THE OPEC KILLER.

STATE FUEL TAX EXEMPTIONS*

Arkansas   9.5¢ applies only to locally produced alcohol or alcohol from states with reciprocal tax credits on Arkansas alcohol

Colorado   5¢ Colorado-produced alcohol only

Connecticut   1¢ reduction

Indiana   4% sales tax on wholesale gas price removed

Iowa   10¢ reduction, increase of 3% sales tax net 7¢ at $1.00 gas

Kansas   5¢ through June 30, 1980; 4¢ through June 30, 1981; 3¢ through June 30, 1982; 2¢ through June 30, 1983; 1¢ through June 30, 1984

Louisiana   all tax removed if alcohol produced in Louisiana

Maryland   1¢ reduction

Montana   6¢ through March 31, 1985; 4¢ through March 31, 1987; 2¢ after April 1, 1987; Montana-produced alcohol only

Nebraska   5¢ reduction

New Hampshire   5¢ reduction on New Hampshire produced alcohol only

North Dakota   4¢ reduction

Oklahoma   6½¢ reduction

South Carolina   4¢ reduction until July 1, 1985; then 2¢ reduction until July 1, 1987

South Dakota   4¢ reduction

Utah   4¢ reduction on Utah-produced alcohol only

Wyoming   4¢ reduction

---

*As of March 3, 1980.

Gasohol, from an economic point of view, is at the crossroads. We are waiting and the price of a barrel of oil goes up a few dollars every month, and the price of gasoline at the pump goes up a dime at regular intervals. We see farm surplus produce piled in mountains, and find it hard to believe that we would starve if motor fuel were to become a farm crop.

The first and last objection to alcohol fuel — its price — is almost certain to vanish in the all-too-near future.

CHAPTER 3

Methanol: Practical Considerations and Precautions

If critics of modern technology distrust the speed of its progress, they should find consolation in that it often goes in circles. A century ago, almost all United States steam transportation was powered by a renewable fuel — wood. For more than a decade, there has been talk, research, and action that will inevitably lead to partial return to wood as the raw material of fuel for the automobile. Muncipal rubbish, which is very difficult to give away, is rich enough in paper (wood pulp) to be converted into methanol, a motor fuel of first quality.

Methanol* is the international chemical name for wood alcohol, or methylated spirits. It is a widely used solvent and raw material in the chemical and plastics industries. It is the low-priced antifreeze — poisonous, like gasoline, but more often hazardous because it may be confused with ethanol, or "grain" alcohol, and drunk. It is practically odorless. It can be made from coal, wood, waste, or any material containing carbon, but, like many other commodities, it is presently made from the most economical source, natural gas. It should be handled like gasoline, although it is somewhat less hazardous. In an engine it burns cleanly, without depositing carbon. It is the only fuel that a wise yachtsman will use in his galley range, as its exhaust is only water vapor and carbon dioxide, identical with the yachtsman's own exhalation. If there is a small fire, a pan of water will extinguish it, not spread it, as occurs with a kerosene or gasoline fire.

---

*Methanol, or CH_3OH

23

METHANOL-GASOLINE BLENDS

Up to 15 percent methanol can be added to gasoline in current cars, without adjustment of the engine, and with noticeable improvement in exhaust quality, economy and performance. Methanol has an octane rating of 106, compared to typical gasolines of 85 to 100. It prevents knocking or "pinging" common with unleaded fuels, and it alleviates "running on" or "dieseling" when the ignition is switched off. These practical and well-documented qualities, added to the virtue of reducing national fuel dependence on the Organization of Petroleum Exporting Countries, have made methanol the leading candidate for the motor fuel of the immediate as well as the foreseeable future.

Methanol blends have been tested in many racing automobiles in the past. A series of tests in stock cars was undertaken by T.B. Reed and R.M. Lerner and their colleagues at Massachusetts Institute of Technology. Nine cars, vintages 1966 to 1972, with horsepower from 57 to 335, were tested on blends of 5 to 30 percent methanol. The cars were unmodified and tests were over a fixed course under standard conditions. A summary* of findings was that (1) fuel economy increased by 5 to 13 percent; (2) carbon monoxide emissions decreased by 14 to 72 percent; (3) exhaust temperatures decreased by 1 to 9 percent; (4) acceleration increased up to 7 percent. The elimination of knocking and of "dieseling" was noted, even on the lowest 5 percent methanol blend tested. The latter improvements were unexpected, but were explained tentatively by the possible dissociation of methanol in the car's cylinder, with attendant absorption of heat energy, quenching early combustion. Simply stated, it burns "cool."

There are problems in the storage and dispensing of mixtures of methanol and gasoline. Gasoline containing 10 percent methanol will absorb 0.1 percent water — ten times as much as gasoline alone. Thus, in a system using the blend continuously, normal amounts of water formed by condensation are carried away to the engine — the "dri-gas" effect.

However, in wholesale storage and distribution of gasoline, water is sometimes used to displace gasoline to prevent the possibility of vapor accumulation and explosion. Residues of this water, with normal leakage and condensation in tanks, are easily separated by traps in transferring gasoline. But when 10 percent methanol is present, the water will desorb the methanol in large amounts.

At freezing temperatures (0° C.), less than 10 percent methanol is soluble in some gasolines. However, it would appear that changes in the

---

*"Improved performance of Internal Combustion Engines Using 5-20% Methanol." R.M. Lerner et al., Lincoln Laboratory, Massachusetts Institute of Technology.

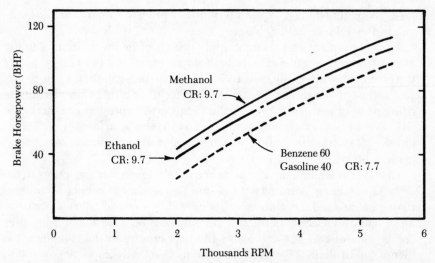

Figure 3-1. In properly designed engines, methanol and ethanol produce more brake horsepower (BHP) than gasoline. A gasoline-benzene blend, used in an engine with a compression ratio (C.R.) of 7.7, produced less horsepower than either methanol or ethanol used in an engine with a CR of 9.7 (SOURCE: Donald A. Howes)

handling of bulk fuel or in the point at which blending occurs would solve the water problems.

The separation of the components may be prevented by the presence of small amounts of higher alcohols in methanol fuels. Curiously, it is easier and thus less expensive to produce methanol with these other alcohols— ethanol, propanol and isobutanol—in it, and the output of a manufacturing plant is increased by 50 percent.

Add Methanol Gradually

A second warning when using a blend for the first time: Add enough methanol to the tank, when taking a trip, to make a 5 percent blend. This will clean out any water in the fuel system. After ten miles, if there is no sputtering, add enough more to make a 10 percent blend.

Drivers wishing to try methanol will be faced with a more difficult problem than cleaning out their fuel system. That is, the problem of finding it in the market. Wholesalers of industrial chemicals and solvents customarily supply it to regular customers in 55-gallon drums as a minimum, or in bulk tank trucks or railroad cars. The supply has always been short, because overproduction would cause storage problems. So unless you know a regular customer "on allocation," who may let you have a few gallons, you will be forced to pay a prohibitive price (for motor fuel) for one-gallon tins sold as shellac thinner, stove fuel, or "spirit solvent."

Large boatyards and marine supply dealers often buy methanol to use as antifreeze in engines in winter storage, or as fuel for yachts' galley stoves. For the latter, the stuff is decanted into one-gallon tins, for which you will pay from three to six times the price of gasoline. Considering the rising price of petroleum products, an enterprising producer of methanol will no doubt seize the opportunity to expand sales by arranging with a retail service station chain, independent of oil company connections, to dispense a premium fuel to motorists.

Critics of methanol blends, often retired engineers writing letters to the newspapers, enjoy pointing out that the alcohols have lower heat contents than gasoline, and therefore must reduce the power and mileage performance. They are correct in the first premise; but because alcohols improve combustion of gasoline and lower the temperature in the cylinders, the blends (up to about 20 percent for methanol) actually improve power and economy. This kind of synergism prevails frequently in the realm of biology, but is seldom perceived or comprehended in mechanics.

Methanol is not highly toxic, but 30 to 100 cc. can be lethal if ingested. It is less dangerous than gasoline if inhaled,* and far less toxic than the two popular household cleaning fluids, trichloroethylene and carbon tetrachloride. If it came into general use, its chief hazards would be controlled by label warnings (not mentioning the word alcohol) and not siphoning fuel by mouth, as is sometimes done in emergencies with gasoline.

Another unsuspected hazard is that of carrying a leaking can in an automobile. Being odorless, the leaking vapor might not be detected in time to avoid considerable inhalation and to avert tragedy.

Advantages of Methanol

There are several advantages to using methanol. A car operated with unleaded gasoline sometimes knocks badly on acceleration. But when a gallon of methanol is added to nine gallons of gasoline in the tank, the knocking disappears. That means increased power, which usually translates into increased mileage. Whether mileage increases 2 percent or 5 percent is not so significant as the 10 percent reduction in petroleum consumption.

Tests with Volkswagens indicate that methanol-gasoline blends low-

---

*It is not known exactly how much, or at what concentrations, methanol can be inhaled without harm.

ered exhaust emissions significantly. With 100 percent methanol, gradual additions of water brought reductions in nitrogen oxides and up to 40 percent fewer aldehydes, another potential pollutant.

Vapor Lock

Some critics complain that there is an increased possibility of vapor-lock. This unpleasant accident occurs in very warm weather with inadequate cooling of a vehicle's fuel pipe by the air stream. Placed under reduced atmospheric pressure by the suction of the fuel pump, the fuel vaporizes. This slug of vapor in the line blocks the liquid fuel, and the driver is forced to wait until natural cooling and condensation occur or to look for a pan of cold water.

Gasoline refineries already seasonally change the characteristic spectrum of volatile components in their products to reduce the incidence of vapor lock in summer and to facilitate starting in winter. Methanol users have reported no problems of this nature. The problem, for either gasoline or blend users, is one that the manufacturers know how to solve, by fuel pump or piping relocation. They refrain from action for the sake of economy.

Corrosion Problems

One possible complaint against methanol as a blender that arose in the 1930s was that methanol was corrosive to certain materials in a car's fuel system. At that time, carburetor floats of cork and gaskets sealed with shellac were easy game for alcohol. Present metal floats and synthetic cements resist the solvent action of alcohol. Carburetor parts are made of zinc die castings, sometimes aluminum. The impurities in both metals in earlier days were conducive to "intergranular crystallization" as a result of aging. This crumbling destruction could be accelerated by the presence of alcohol and water, but the problem no longer exists. Lead, tin and magnesium are attacked by methanol, but there should be no opportunity of exposure to these metals in the combustion zones of an engine. Iron and steel are quite immune, as are brass and bronze.

Users of pure methanol found an unsuspected cause of trouble in the gasoline tank, which traditionally has been made of "terne plate," a favorite roofing material of Victorian architects. It is steel sheet coated with lead, making it ideal for resisting rust from water in gas tanks. Methanol reacts with lead, slowly but surely, forming a flaky sludge that plugs filters in fuel pipes. The easiest solution is to inspect and clean the filters every few days when starting to use methanol. The lead will all be gone in a week or two.

The more rational solution will be the decision of manufacturers to abandon terne plate for an epoxy-coated lining in a plain steel tank. The first solution is described in Appendix A. (See Appendix A, the Mx-100 test car.)

100 PERCENT METHANOL FUEL

North American and European ventures into alcohol fuels, both in recent times and in the 1930s, depended on blends ranging from 2 to 95 percent alcohol. The motivations were numerous, but the underlying reason for adhering to blends was that no alteration of the automobile engine was required. However, about 1910, pure alcohol was an established alternative fuel for the horseless carriage. Assuming the availability of methanol at the service station around the corner, what kind of engine would be designed for its maximum benefits?

First, the amount of air consumed by burning methanol is reduced. While the ratio for gasoline is 14 weight units of gasoline to 1 of air, for pure methanol it becomes 6 parts fuel to 1 of air. Carburetor jets must be changed.

Second, heat is needed at the methanol intake to vaporize it. A loop of exhaust pipe around the air intake will do the job.

Third, an initial cold start requires an electric glow-plug or a highly volatile primer fluid, such as gasoline or LPG (Liquid Petroleum Gas). Priming fuel is fed to an auxiliary valve from a small tank. Such a conversion might cost around $100, if made on an existing car. If incorporated in a production model, the cost would be one-tenth, or nothing, as the trade-off could be the elimination of emissions control devices.

Methanol burns without misfiring at leaner mixture ratios than gasoline. This does not mean that it uses less fuel — in fact, methanol quantity consumed will double — but it indicates more complete combustion and less pollutant exhaust than gasoline. Temperature of exhaust is 100 degrees centigrade lower than with gasoline, and spark timing can be later (more efficient) with methanol because of its higher flame speed. Higher compression ratios are feasible without toxic additives, and "dieseling," or running on after the ignition switch is turned off, is eliminated because of the higher heat of methanol evaporation. (Methanol is not a good diesel fuel.)

For the ordinary driver, these qualities will be apparent in improved acceleration, a lighter car and engine, fewer parts to service (no catalytic converter, etc.), but a car that needs about twice the fuel-tank size for the same mileage as gasoline. This problem caused the sad failure of the *Novis*, an advanced design of Indianapolis "500" race entries, designed to run on 100 percent methanol but unable to finish the race at the speeds of their qualifying runs. The rules limited the fuel used—in gallons instead of in heat units.

Recent Test Results

A 1978 Ford Pinto station wagon has been driving around Sacramento and Santa Clara, California, using only methanol in its tank. The modifications to the engine, and the addition of a fuel injector and instrumentation, were made in a period of two months with a budget of $50,000, a small amount for this type of experimental work. Similar stock-model, low-priced, mass-produced Volkswagens and Volvos have been driven in Germany and Sweden. The cars have been used to test various methanol fuels and accessories. Car performance, maintenance, durability and other factors that might be influenced by running on methanol were also examined.

A Mazda Mizer sedan and a Datsun truck modified at Texas A & M College to use and test methanol, have logged over 20,000 miles since 1975. Daimler-Benz engineers published data on tests of a Mercedes-Benz

Volkswagen researchers have put about 1 million miles on a test fleet of VWs and Audis powered by a mix of 85 percent gasoline and 15 percent methanol.

4.5 liter V-8 car, running on methanol, in 1975. In this test, methanol produced lower emissions than the usual "super" fuel (gasoline and lead), as shown in the accompanying table.

Fuel	Carbon Monoxide grams/test	Hydrocarbons grams/test	Nitrogen Oxides (NO_x) grams/test
European legal limits* (after 10/1/75)	149	9.7	—
"Super" fuel	140	6.0	8.0
Methanol	32	5.5	0.72

*European legal limits are approximately equivalent to U.S. Department of Energy and Environmental Protection Agency emission standards.

The test did not include the emission of aldehydes, an unpleasant product of combustion of alcohol resembling formaldehyde. There is not enough experience, thus far, to indicate whether they are harmful, beyond the slight irritation caused to the eyes and nose by prolonged proximity. But one very obvious advantage of pure alcohol fuel is seldom mentioned: water can be mixed with it, either in the tank, or by spraying it into the intake manifold. Also, it may be injected at times when maximum power is needed. Water reduces combustion temperatures, thus reducing nitrogen oxide (NO_x) emissions and increasing power through higher, knockless compression. Water should not be mixed intentionally in methanol blends, but injection, by a separate device, is possible.

While the A & M testing was done with carburetors or vaporizers and air induction of the methanol, the most recent California Pinto tests, as well as those of Mercedes, used engines with fuel injection. The complete data on the Pinto modification and test results appear in Appendix A.

The increasing use of fuel injection for small economy cars is significant for the straight methanol cars of the future. Fuel injection provides more reliable starting, better economy, leaner combustion (perfect air/fuel ratios) and hence less emission of pollutants than carburetion. Injection becomes even more desirable with methanol because of its high latent heat and its high surface tension. These are the characteristics that make it difficult to vaporize in a regular carburetor but make it a smoother, cooler fuel when once in the cylinder. Another advantage of fuel injection is that it is easily and accurately controlled by an electronic "black box" that senses all the physical environmental and engine conditions and coordinates them to modify fuel input and ignition timing.

COMPARISON OF COSTS

Before the discovery that automobile air pollution had an intimate and enormously expensive relationship with public health, the price structure of a fuel system was the sole criterion of its validity. A pre-smog era, mathematical approach to methanol's chances of success in the old, unrestricted market would begin with an analysis of its heat content and comparative fuel consumption with gasoline:

Fuel	Consumption Ratio
Gasoline	100.0 (Base)
Ethanol	161.4 Volume to equal gasoline
Methanol	221.6 Volume to equal gasoline

If we took these data at face value, methanol would have to cost less than 100/221 times the price of gasoline to be competitive. For example, if gasoline were at $1.40 per gallon, methanol would have to be below 63.2 cents per gallon to be competitive. Although this price is considered reasonable by some even now, we cannot expect to have cars that burn pure methanol for some time. Hence, an example using blended fuel is more pertinent.

Assume a 10 percent blend of methanol:

Cost of 9 gallons of gasoline @ $1.20/gallon	$10.80
Cost of 1 gallon of methanol @ $1.50/gallon	$ 1.50
Total cost of blend	$12.30
Cost of 10 gallons of gasoline @ $1.20/gallon	$12.00
Additional cost for blend	$ .30

Assume that the car goes 20 mpg on gasoline, and has a 6 percent increase in mileage on the blend, giving 21.2 mpg (a conservative figure). On the blended tankful, the car goes 212 miles, against 200 on gasoline. Since it costs 6¢ per mile to drive this vehicle, 12 miles gives a credit of 72 cents, so the saving of the blend is 72 minus 30 = 42 cents, or a 42-cent bonus for using an expensive additive!

The higher latent heat of alcohols reduces intake temperature of the fuel-air mixture, compared to gasoline. Hence there is higher volumetric efficiency for any given horsepower output for blends up to 20 percent alcohol. Richer alcohol contents increase fuel consumption at rates that are more in proportion to the heat content.

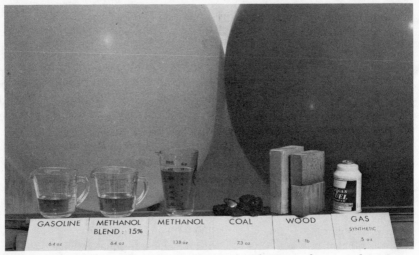

This photograph shows the amounts of several fuels required to propel for 1 mile a car that goes 20 miles on a gallon of gasoline.

This nonlinear effect of alcohol additives has made possible the surprising mileage improvements reported by Reed and Lerner and has explained reductions in toxic emissions far beyond the 10 to 20 percent alcohol contents involved. The message of this phenomenon is simple and heartening: *The greatest value of methanol, as a transitional additive and extender of gasoline, occurs in the range of proportions that requires no change, and little adjustment, to present cars.* These facts were established first by E. Hubendick* and reported to the World Power Conference in 1928.

METHANOL AND SYNFUELS

Although methanol was first found in nature (it was derived from the destructive distillation of wood, during the manufacture of charcoal), the modern product is entirely synthetic. That is, its molecule is put together by combining hydrogen, carbon and oxygen in the presence of a catalyst.** Gasoline consists of a large assortment of natural hydrocarbons distilled from crude oil, although it may be improved by cracking to elevate the octane rating.

---

*Alfred W. Nash and Donald A. Howes, *The Principles of Motor Fuel Preparation and Application* (New York: John Wiley & Sons, Inc., 1935). Vol. I, "Alcohol Fuels," pp. 349-451; Vol. II, "Motor Fuel Specifications."

**$C + H_2O = CO + H_2 = CH_3OH$

When President Carter made his nationally televised energy speech of July 15, 1979, he used the word *synfuel* without adequate definition. His meaning, on which he had been coached by representatives of the oil industry, was synthetic gasoline and oil. Such products, made from coal, were seized upon by Hitler as the only way to political and military independence. Four years before World War I started, Friedrich Bergius discovered a way to add an atom of hydrogen to coal — hydrogenation — and produced synthetic oil. Before World War II, the Fischer-Tropsch and Lurgi processes improved over that of Bergius. After the war, Krupp-Koppers entered the competition, selling over 40 coal gasification plants to developing countries. The main product of these was fertilizer.

Hitler's needs in 1941 were for aviation gasoline and industrial oil fuels, and they both had to be compatible with the natural petroleum that Germany was able to obtain from its allies. The demand for interchangeability ruled out straight methanol for military vehicles, but it was blended with gasoline, as ethanol is now blended with gasoline to produce Gasohol.

There is no such emergency today in the United States to force a commitment to making synthetic gasoline, particularly when its cost is guaranteed to be higher than that of the natural product. Mobil Oil has developed a synthetic gasoline in a process that goes from gasified coal to methanol and then to "syngasoline." This approach, researched at government expense, is often questioned by taxpayers familiar with the superiority of methanol fuels. Why double the cost by converting a good fuel to one that has caused endless problems in pollution, efficiency of engines, and performance?

A subsidiary of Gulf Oil is running a pilot plant at Fort Lewis, Washington, that makes 100 barrels of oil a day from 30 tons of coal by dissolving it and adding hydrogen at high temperature and pressure. Sulfur and ash are separated from the oil by distillation. Hence the process creates no air pollution. However, ecologists are quick to deplore the huge gashes in the landscape that accompany strip mining. A plan to build the country's largest coal gasification plant near Beulah, North Dakota, is raising the ire of farmers in the area, who ask what will happen when the seams of coal (lignite) run out in a few decades.

Other coal liquification processes are being developed for pilot plant demonstrations by Ashland Oil in Kentucky and by Exxon in Texas. Although the larger of these, Ashland, will handle only 600 tons of coal a day, it will all be observed carefully for economic efficiency. The Gulf plant at Fort Lewis is thought to be the most advanced in technology, and has been the basis of planning a plant in Morgantown, West Virginia. Construction will take from 1981 until 1984. Capacity will be 20,000 barrels of oil per day from 6,000 tons of coal. This is a small plant, from the production viewpoint, but if it demonstrates the feasibility of the process, it will be scaled up to 100,000 barrels per day. International interest in this plant is keen. West Germany's government will share a fourth of the cost, with Gulf and possibly the Japanese government sharing the balance with the Department of Energy. The total cost will be over $700 million.

Objections to spending the $100 billion or more in the coming decade on synfuel plants, programs for which the above experiments are merely toy-models, are coming from every quarter except the oil industry. Environmentalists say that synfuels will cause heating of the atmosphere, (through carbon dioxide pollution) with a resulting rise in ocean water levels as the polar ice caps melt, destruction of agricultural lands, and desecration of vast areas of scenic landscape by open mining. But the first of these threats to the world's life would occur if we continue to burn fuel of any kind at present rates. Hence conservation should come first in all programs for reduction of energy use.

Fuels from renewable sources, mainly methanol and ethanol, must be produced as rapidly as possible because they are compatible with gasoline, for the period of transition. Last, because they require more capital than is now forthcoming, would be synthetic fuel plants to supply industry and the military with heavy oil and distillates.*

The primary concern of conservationists, beyond their immediate problems in gasoline supply, should be the insulation of buildings. Expansion of public transportation facilities ranks next. Then follows the use of municipal, agricultural and forest wastes for motor fuel. The daily collection of 10,000 tons of garbage in New York City might convert to 3,000 tons of methanol, which could be used "straight" in fleet vehicles, or become the additive for 10 million gallons of Gasohol.

METHANOL PROJECTS

Maine has 5.5 million acres of forests afflicted by spruce budworm that is killing the trees, and a methanol plant was considered for converting this carbon supply to liquid fuel. Lack of funding is the principal deterrent to this project.

In addition, all of Maine's forests could yield wastes that would supply up to a billion gallons of methanol each year, an amount matching the present total United States annual production. Although it is not as effi-

---

*Fuel used by the armed forces is 3 percent gasoline; the balance is jet fuel, diesel, and bunker fuels.

This is a Morbark machine harvesting spruce in Maine's experimental cleanup operation.

cient to produce methanol from wood as from coal — wood-produced fuel might cost around 50 cents per gallon, compared to about 28 cents from coal — the combined salvage nature of the project and the relative isolation of Maine from distribution centers of gasoline are incentives for local methanol plants.

An even better incentive appears when combustible municipal wastes, that normally cost $3 to $9 per ton for disposal, are delivered to methanol conversion. Americans use about 575 pounds of paper per capita each year, and this waste could, with other solid wastes, supply about 8 percent of the transportation fuels we need, if converted to methanol. The chief deterrent to wood fuel utilization is that the world's forests are at the greatest possible distances from the centers of population (and utilization), but urban methanol plants place the material source, refuse, and the fuel market in immediate contact.

The abatement of federal and state taxes on methanol fuels would certainly encourage investments in methanol. Educational campaigns through the Environmental Protection Agency, the Department of the Interior, and state conservation groups could be effective. However, the gravest impediment to the free development of methanol as an alternative fuel is that it would be almost impossible to gain a proprietary hold over its raw material sources. Carbon is everywhere, and a monopoly on rubbish, wood chips, moldy corn crops, or marginal coal fields, is not as attractive as control over a nation's oil industry, from well to gas tank.

However, as petroleum prices rise, there will surely come the occasion

35

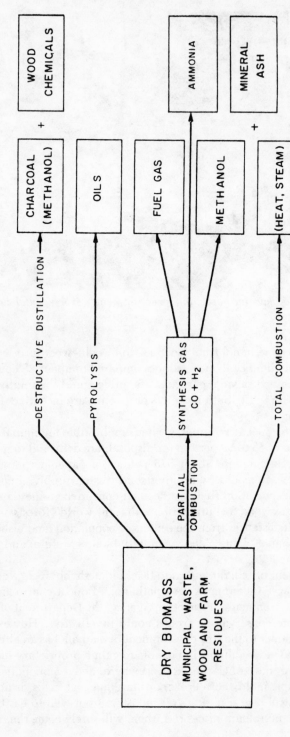

Figure 3-2. Methanol is one of many products obtainable from dry biomass.

to reexamine the economies of methanol. In Arabia, millions of cubic feet of natural gas are flared off (burned at the wellhead). This gas is a prime source of methanol, which can be shipped and stored at a fraction of the cost of liquified natural gas (LNG) to the thirsty engines of the world. Costs at the refinery might be less than 15¢ per gallon. The construction of Arabian methanol plants would enable them to undersell our methanol, and maintain an even higher trade deficit for the United States.

THE FUTURE OF ALCOHOL

Conversion to a complete alcohol fuel economy, with methanol playing an important role, will not come before late in this century. The resistance to any upheaval has been formidable in the past. The automobile industry, asked by government agencies to make cars safer, cleaner, and more economical, still drags its feet, requesting extensions of the time limits and complaining of the costs. An exception was the about-face of General Motors. On August 1, 1979, Elliot Estes, the GM president, announced that the firm would exceed the mileage requirements set by the government before 1985.

In 1975, Congress unknowingly threw a wrench into the pure alcohol machine, with its mandate to the automobile industry to raise their fleet average of miles per gallon from 14 to 27.5 by 1985. This places an almost insurmountable penalty on new cars designed for alcohol. The volume consumption of pure methanol (and ethanol) fuel is sure to exceed gasoline, from 30 to 100 percent, but methanol will be cheaper, and it will be made in the United States.

Amend the Law

The law should be amended so that cars made to run on straight alcohol would be excluded from the miles-per-gallon requirement, inasmuch as their consumption of domestically produced nonfossil fuel costs us nothing in foreign trade deficits and nothing in pollution regulation or remediation. The secretary of the Department of Transportation must carry out the congressional mileage mandate on a schedule that makes the manufacturers wince. The greatest improvement is required in the early 1980s, tapering off by 1985. The automakers would rather have it the other way around.

If tackled much earlier, the mileage controversy might have been settled by following the example of the English. They taxed the big "guzzlers" on the basis of horsepower. The resulting taxes on large cars made the British a nation of tiny cars with sporty but economical engines.

American manufacturers are working feverishly to achieve the 27.5 miles per gallon average by 1985, which means some cars must get well over 30 mpg to compensate for those luxury cars in the teens. But they are temporarily obliged to lower compression ratios. This counters the long-range need for lighter, more fuel-efficient cars that result from high-compression engines. The reason for this perversity?

The high-compression engines of 1973 and earlier tend to knock if run on unleaded gasoline, as they are required to do to avoid ruining their catalytic converters. Lead toxicity is the target of environmentalists' plans to eliminate it from all fuel by 1985. Fortunately for owners of the pre-1973, efficient, high-compression cars, alcohol, in the proportion supplied in Gasohol, is an excellent octane booster. It would make the older cars run cleaner, more efficiently and without lead, but there is not enough Gasohol to reach the drivers who need it most.

Foreigners Are Leading

Research on alcohol fuels and engines that use them effectively has been conducted since the end of World War I in Britain and Europe. The key reasons: a desire for relief from the economic bondage of imported petroleum and a consuming interest in motor racing and high-performance sports cars. Tests performed in the 1930s in England show the increase in output of a 2-liter engine running on methanol. Even with the moderate compression ratio (9.7) the improvement was significant. Modern testing instrumentation and cumulative combustion technology have made CR's of 14 practical, as in the California Mx-100 (Appendix A).

If the Detroit automakers plan for future straight alcohol cars, as they will be forced to, their timetable will probably take the course of all of the past major innovations, such as four-wheel brakes, overhead cams, and front-wheel drive. These came in from Europe, and gained a market foothold before Detroit followed, usually after ten years. I predict that the first all-alcohol cars to enter the market here will be Japanese sports cars.

METHANOL AND ETHANOL RELATIONSHIPS

Ethanol will eventually be an important, if not exclusive fuel in parts of the United States, Brazil and other countries. However, methanol has an equal if not superior chance of growth as a fuel.

In the past, ethanol has cost two to six times more than methanol to produce. The difference is attributed to higher costs of raw materials, the inefficiency of the batch method of fermenting and distilling used in conventional ethanol plants, and the time required in the process.

The synthesis of methanol, on the other hand, is a recent technology, in which a continuous input of gas to the equipment produces a stream of methanol. As long as natural gas is used as the raw material for methanol, and as long as gas is available at reasonable prices, methanol can undersell ethanol on a volume and energy basis.

At present, the range of raw materials for methanol manufacture is wider than for ethanol, as it includes cellulose (wood and paper wastes) which cannot be converted directly to ethanol. However, in several laboratories, work is being done to improve the known methods of converting cellulose to glucose by enzyme action. This would make it possible to produce ethanol from municipal and farm wastes that otherwise supply only solid fuel for conventional heating and steam generation. Despite the claims of inefficiency by oil industry scoffers, ethanol is being produced with very low levels of heat, from solar sources and in equipment collected entirely in junk yards. It is an "intermediate technology," in the words of E.F. Schumacher.

With comparable diligence, there are inventors busy with new ways of producing methanol, particularly from biomass wastes and from coal. These developers are searching for ways to reduce the high temperatures and pressures, and the attendant costs, now associated with mass production of methanol. Their ideal would be a portable machine that could sit in the corner of a farmer's barn and produce tractor fuel, on demand, from corn stalks, straw or wood chips.

Research scientists at the Solar Energy Research Institute at Golden, Colorado, are working on a gasifier that they hope will scale down the present concepts of methanol and ammonia synthesis plants to sizes compatible with today's agricultural production. The high yields of corn that make the Gasohol movement practical depend on steady application of fertilizers, which, in turn, depend on the production of ammonia. As farmers are well aware, ammonia is produced mainly in plants fueled by petroleum, and critics of the ethanol fuel program are quick to see the folly of using oil to save oil. The versatile methanol-ammonia plant will solve the problem, because the plants to produce each commodity are almost identical.

The synergistic triangle of ethanol, ammonia, and methanol begins to emerge as a logical fuel, food, and fertilizer chain for the agricultural states. The feasibility of portable ammonia plants was demonstrated during World War II, when the government averted both shortages of fertilizer and ammunition by prefabricating ammonia plants on skids. It is possible to visualize a self-sufficient farm community, pooling its resources of surplus biomass, operating a versatile synthesis plant to produce fertilizer during the winter and motor fuels during the growing seasons.

Ethanol as a Mixing Agent

Both ethanol and methanol are needed for future automobiles. In northern Europe and Scandinavia, where the primary alternative fuel is expected to be methanol, and where winter temperatures drop to arctic lows, the initial blended fuel, comparable to our Gasohol, will be 10 percent methanol with a mixing agent, probably ethanol. The need for this agent, when methanol is concerned, is explained as follows:

When very small amounts of water accumulate in a fuel tank containing an alcohol-gasoline blend, the two components will tend to stratify as the temperature drops. This problem, mentioned earlier, is called *phase separation* (see Figure 3-3). The alcohol settles at the bottom of the tank. When it is pumped up for a cold start in the morning, it fails to vaporize in the carburetor, and it is necessary to apply heat to the fuel for an initial start. This is why the production of Gasohol demands that the ethanol be of 200 proof, and *anhydrous* — without water. With that purity, it will stay mixed.

The equipment for the heating function is available and will be a routine accessory on cars designed to burn 100 percent alcohol, but it should not be necessary for the use of blends such as Gasohol. But even if our hypothetical car in the arctic started, it would not run properly until all the alcohol at the bottom of the tank was used and gasoline finally reached the carburetor, which is adjusted to burn it.

Methanol is much more sensitive to separation problems due to moisture than ethanol. Fortunately, the synthesis of methanol normally produces "methanol fuel," containing small but significant percentages of the higher alcohols, isopropyl and butyl alcohols in particular, which are important to the miscibility (ability to mix) of the methanol with gasoline. The higher alcohols are removed to produce pure methanol, so that the price of the "impure" fuel can be lower than that for pure methanol. Another advantage of the raw fuel production is a higher yield per unit of input energy. Figure 3-3 shows the importance of the higher alcohols and ethanol in preventing phase separation at zero degrees centigrade (32° F.), and similar data for a temperature of − 10° C. (+ 14° F.). Ethanol will play an important role as a mixing agent, as it is almost impossible to eliminate water from the car's environment.

Stratification has not appeared to be a serious problem with those testing methanol blends from Massachusetts to California, even at freezing

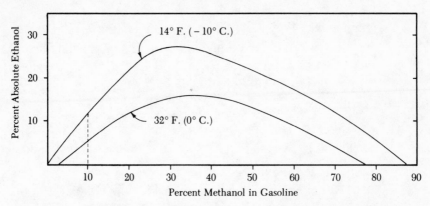

Figure 3-3. Phase separation (stratification) of methanol-gasoline blends, at freezing temperatures, may begin with the blend contains about 4 percent methanol. With 10 percent methanol and 90 percent gasoline (dotted line), 5 percent ethanol will prevent the blend from separating at 30° F. At 14° F., the addition of 10.3 percent (200) proof ethanol is required to prevent separation. (SOURCE: Howes and Institution of Petroleum Technologists)

temperatures in the former state, possibly because of the low percentages (5 to 10) of methanol and the movement of cars that tends to mix the fluids. However, it would appear from the laboratory data on miscibility of methanol-gasoline blends that motorists in subfreezing areas, if and when methanol blends are introduced, would be prudent to avoid stratification by insisting on the presence of one or several of the mixing agents.

Areas already using Gasohol in winter conditions have not reported difficulties attributable to the ethanol component of the fuel. Dealers have been extremely careful to avoid water contamination. Hence, it appears that a gradual change from ethanol to the cheaper methanol component in Gasohol, if it occurs, may be effected by retaining enough ethanol to maintain adequate mixing in winter. The proportions will be determined by regional fuel distribution technologists and experience.

These pages have been devoted mainly to alcohol fuels as a gradual replacement, in volume, of fossil petroleum-based domestic and foreign fuels. Alcohol is the only fuel that can possibly substitute, during a gradual buildup of its production capability, for gasoline, without disrupting the automobile industry or the retail fuel distribution system.

CHAPTER 4

Portable Gas Generators

Soon after the peak petroleum shortages in the 1970s, many vehicles essential for maintenance tasks in municipal services, particularly gas companies' vans, appeared with bumper stickers proclaiming that they were "running on clean fuel." They were equipped with carburetor adapters and accessories enabling them to use a steel bottle of natural gas as their primary fuel source. Except to a perceptive driver, the conversion made no difference in performance of the vehicle and no external, visible change in a car's profile. However, the exhaust is so harmless (water vapor and carbon dioxide) that trucks operating entirely inside warehouses and factories may be so powered.

FACTS ABOUT PORTABLE GAS ENGINES

An infinitesimal fraction of the motoring public is aware of the use of gas as a motor fuel, and of its compatibility with the gasoline engine; accordingly the portable gas-maker has been a total stranger in the United States since 1914. However, it provides a simple, relatively inexpensive means of converting any number of solid fuels to a useful automotive gas, without recourse to the traditional monopolistic systems that produce and dispense petroleum fuels. It makes this fuel while traveling, more or less in proportion to the needs of the engine; it can function on coal, coke, briquets, wood chips, or corn cobs, if nothing else is available.

Less Power/More Care

In exchange for this ready versatility, the gas generator (*gasogen*) demands certain sacrifices, the most noticeable of which is diminished power. Even the smaller U.S. compact cars, however, have considerably

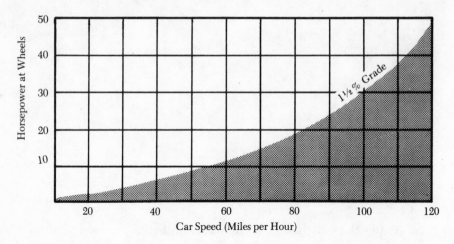

Figure 4-1. Horsepower required to propel a 1,700-pound, well-streamlined car of a drag coefficient of 0.32 at various speeds. Data computed and experienced. Modern radial tires would improve performance.

more power than is considered necessary by unbiased engineers and European economy-car builders.

The second demand, which may only assert itself when it is too late to heed it, is the need for frequent personal attention to renewing filters, to cleaning the purifier system, and to disposing of corrosive and foul-smelling by-products before they have a chance to enter the sensitive confines of the engine's lubrication system.

The loss of power, compared with gasoline or natural gas, occurs because the gasogen uses air to oxidize its carbon fuel, and air is 80 percent nitrogen, which is passed on to the engine as a useless diluter of the real fuel, carbon monoxide. The actual power loss is around 40 percent.

There are known techniques for dealing with both of these disadvantages of on-board fuel generators, but it is obvious that the seriousness of world events necessary to drive a significant segment of vehicle fueling to gasogens would also increase the drivers' tolerance for inconvenience.

The major difficulty, in the event of emergency, would be the immediate acquisition of fuel sources for urban motorists. In the long run, however, the traditional solid fuel sources, coal and wood, provide potential resources for nearly all transportation needs of the foreseeable future, whether produced in transit or converted, with more efficiency, to liquid fuels in large plants.

Heated, Then Cooled

We have mentioned two energy losses in gasogens: nitrogen dilution and the human energy expended in cleaning operations. A third, not as responsive to mechanical remedies, is inherent in the outwardly simple conversion of carbonaceous fuels to combustible gases. This is the loss of heat units occurring when the fuel is burned in the retort, and then cooled before entering an internal combustion engine. The cooling is necessary to clear the gas, by condensation, of acids and tars harmful to the engine, and for volumetric efficiency, as explained later in this chapter.

Although it would be an example of "out of the frying pan, into the fire," this problem could be avoided by changing the engine to an external combustion affair: steam, Stirling cycle, or a gas turbine in which all available heat units are recycled (see Chapter 5).

Gasogen Reactions

Reactions taking place within the gasogen are well known and should be understood by everyone contemplating construction and operation of one (see Figure 4-2, page 46) for reasons of safety, and efficiency in fuel consumption. Although the various zones of chemical activity and temperature were clearly defined for steady state production for city gas supply in manuals of the nineteenth century, there may be wide deviations from the norm in automotive application. For example, the temperature of the combustion zone influences the gas quality drastically, as shown in the accompanying table.

**EFFECT OF COMBUSTION ZONE TEMPERATURE
ON GAS QUALITY**

Temp. Deg. C.	Temp. Deg. F.	Composition of Gas by Volume			Water Vapor Dissociated
		H_2	CO	CO_2	
674	1247	65.2%	4.9%	28.9%	8.8%
861	1570	59.9	18.1	21.8	48.2
1125	2057	50.9	48.5	0.6	99.4

G. Rouyer, Dunod, Paris, 1938, "Etude des Gazogenes Portatifs." (Quality of fuel gas as controlled by temperature of gasogen combustion zone.)

The second most important determinant of gas quality is the fuel itself, if one is to heed the experiments of one Bell, quoted in the earlier Rouyer reference:

"The following figures show the results of passing a stream of CO_2 over three gasogen fuels, all at identical temperatures. The superiority of charcoal as a producer of combustible gas is evidently due to porosity, and suggests the greater efficiency of pelletized fuel and fluidized bed techniques in the future:"

Gas	Dense Coke	Porous Coke	Charcoal
CO_2	94.56%	69.81%	35.3%
CO	5.44%	30.19%	64.7%

Here the superiority of charcoal over coke is evident, with higher carbon monoxide. Bell has neglected to list hydrogen content, which might

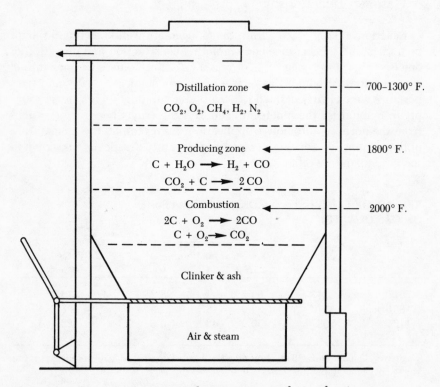

Figure 4-2. Diagram of a nineteenth-century gas producer, showing approximate zones and temperatures. Coal is used as fuel.

have made a slight difference. Although the data are sketchy, we have inherited enough from the 1940s to know that the secret of efficient gasogen operation lies in maintaining a constant load on it (speed of engine) or in automatic and highly sensitive control of temperature, which must, in present designs, be linked directly to combustion air input, and, in turn, to intake manifold vacuum and engine revolutions per minute.

Fortunately for several hundreds of thousands of Europeans during the decade of World War II, gasogens on trucks, cars and tractors performed their tasks despite careless operation and maintenance, and despite extremes of temperature and fuel quality. Much of this adaptability and tolerance is due to the simplicity of the basic generator design.

It's Like a Stove

It is no more than a stove, usually turned upside down, and in its elemental form, the gasogen has no moving parts. At worst, it has a moving grate, and an electric or hand-driven blower to get it started. At the top there is a fuel loading door; in about the middle, there is a small door to light the fire; at the bottom another door, for removal of ashes. Thus far the description matches that of an ordinary old kitchen or laundry coal stove.

The flue, or chimney, is what distinguishes the gasogen from a stove, as it is only four inches or less in diameter, as compared to six for a coal stove. Another difference is that air is admitted *over* the fire, which burns downward, sidewise, or upward, depending on the model. Air is drawn in through small tubes to the center of the firebox, instead of through an opening in a door. The tubes are called *tuyeres*, or pipes, as the French claim credit for this piece of hardware, originally for blast furnaces. However, there is no fireside glow about these pipes, as they are shrouded by an air duct, which directs cool air over the inner tips of the pipes to prevent them from melting, and often connects with a blower, used in starting the fire. A pipe leading directly upward, with a shut-off valve, vents the useless smoke of starting the fire to the atmosphere. Figure 4-3 shows typical construction for a gasogen.

In the days of coal heating of tenements, the first cold snap of winter brought death to many immigrant families, often from the Mediterranean, huddled in a room with a stove. With windows caulked with newspapers, the air supply to the fuel was inadequate. A change in wind direction or the addition of coal caused the draft to subside, and carbon dioxide (normal flue gas) passed downward, over the hot coal, losing an oxygen molecule, and escaping into the room as carbon monoxide — bad news in a stove but the life-blood of an internal combustion engine, whose unhappy

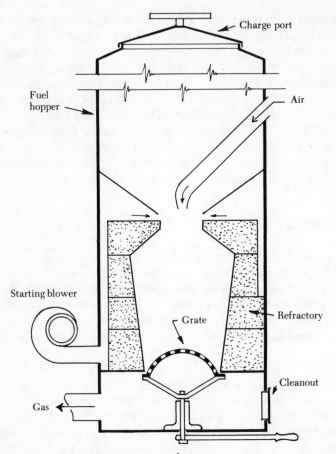

Figure 4-3. An early and heavy gasogen design.

lot it is to be weaned on a gasogen. This gruesome explanation notwithstanding, the hazards of the gasogen are almost negligible, as it must be airtight to work, and the engine inhales the deadly fumes and emits an extremely clean exhaust of carbon dioxide and water vapor.

Gasogen Design: Trial and Error

The design of gasogens, as may be guessed, is by trial and error, and the major variables are usually controllable from the driver's position. For a certain engine at a given cruising speed, with a fuel of known moisture content, size, heating value and density in the firebox (which is invariably combined with an overhead magazine of fuel), there will be an optimum

A factory-built gasogen on a French sedan, installed in the trunk space. Upper curved part stores charcoal, roof rack stores more. On front fenders are the gas cleaners and coolers. This is a typical World War II installation.

draft, or negative pressure, in the firebox, adjusted by a valve. As the CO is sucked toward the engine, air for its combustion must be admitted near the intake manifold. This is adjustment number two. The third one is the valve that connects the normal gasoline carburetor to the engine for starting, and then, when the gasogen's fire is going (a matter of skill and minutes), switches off the gasoline and connects the engine with the gasogen.

Driver Skill Needed

It may be seen that there is plenty of scope for skilled manipulations of these controls, as drivers prepare for a long hill ahead, building up a reserve of heated firebed by shifting down, which increases air intake volume. On the other hand, a long downgrade might allow the fire to go out, unless the engine is "gunned" occasionally. One can now appreciate the relative elegance of modern carburetors that make all these judgments and adjustments automatically.

The largest part of the gasogen's extensive installation is not the retort but the cleaning and cooling equipment necessary to remove tars and acids, (a whole chemistry laboratory-full, if you are indiscreet enough to burn soft coal) and other solids that would contaminate and abrade the engine. French gasogens described in the historical sketch (Chapter 6) were designed to run on lignite, imposing maximum cleaning problems.

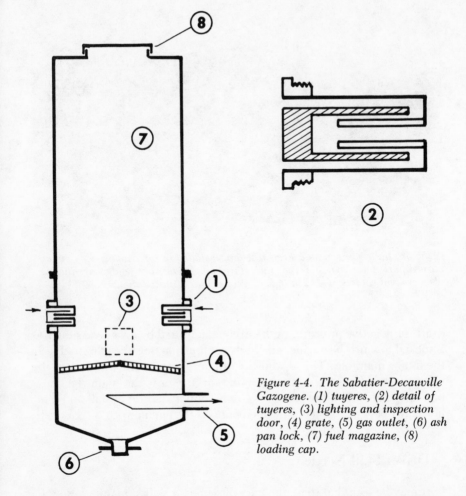

Figure 4-4. *The Sabatier-Decauville Gazogene. (1) tuyeres, (2) detail of tuyeres, (3) lighting and inspection door, (4) grate, (5) gas outlet, (6) ash pan lock, (7) fuel magazine, (8) loading cap.*

Charcoal and coke, having been cooked until most of their volatile contents have departed, need less attention to cleaning than other fuels. However, their activity and temperature of burning are high, and a large cooling area must be considered.

If the internal combustion engine is fed on air that is hot (i.e., expanded), it cannot gulp as much of it as if the air were cool. The gas from a gasogen is dilute enough — with carbon dioxide and nitrogen — without further loss by heated air intake. This is one reason for coolers of good size (volumetric efficiency). The other reason, equally important, is that the lower the temperature, the less tar gets into the engine, and the more of it sticks in the sumps of filters and collectors.

MANY FILTER MATERIALS

A list of the materials that have been used for filter beds would fill a page. Some materials, like ceramic rings, stone chips, and metal gauze, are permanent, and must be washed off periodically. Others, like mineral wool, jute or hemp fibers, and wood chips, are disposable (better be cautious). These necessary, frequent service jobs are the real complaint against gasogens; and in a well-designed outfit, quick-release latches and accessible replacement filters will be included. The British Ministry of Transport designs for its equipment were excellent in this respect, as they imitated vacuum cleaner fittings that any child can manipulate.

The only part of a gasogen outfit not readily made by the simplest sheet metal and welding techniques is the changeover valve, or carburetor shunt. It is similar to that supplied by dealers in propane conversions, but much larger, as propane gas is *all* fuel and rather concentrated. Another essential, easily purchased, is a gasoline shut-off valve, with electric actuation. This should be switched to the closed position automatically when the gas valve is opened, the carburetor shunted, and the engine begins to run on carbon monoxide gas.

When the supply of gasoline is strictly curtailed, or when it is at such a distance from the user's base that it is inconvenient or expensive to fill one's tank, the annoyances of the gasogen diminish. They are also reduced if fleets of vehicles are equipped with gasogens and maintenance is done routinely by specialists.

Sports cars and racing cars, particularly in European practice, used the positive displacement "blower" or supercharger to increase the amount of fuel-air mixture entering the cylinders. Increases in maximum horsepower and acceleration performance varied, but they might rise as high as 25 percent so long as the increased compression ratio did not cause excessive knock. Modern engines are pushing the limit closely, and superchanging thus seems an unprofitable device for increasing the power of generated gas.

ADDITION OF OXYGEN

Oxygen, once an expensive industrial luxury for certain processes where nitrogen is undesirable, is now sufficiently low-priced to be considered for the combustion of subterranean coal and of municipal wastes. A steel bottle of oxygen, piped through pressure regulators to the gasogen and sup-

COMPARISON OF FUEL CONSUMPTION
IN SVEDLUND GASOGEN

Fuel	Water Content	Equivalent in Gasoline
2.8 kg. wood (6.1 lb.)	18%	1 liter
23.3 lb. wood	18%	1 U.S. gallon
1.4 kg. charcoal (3.08 lb.)	7%	1 liter
11.65 lb. charcoal	7%	1 U.S. gallon

plied for acceleration and hill-climbing by automatic barometric sensors in the intake manifold, might restore a large part of the power lost through nitrogen (air) dilution, excess carbon dioxide, and low temperature of the reaction zones.

The reactions involved become more favorable to fuel quality as the temperature is elevated; hence, the control of this parameter, as well as internal pressure (normally negative) would surely improve performance. Solid-state electronic controls make this kind of improvement, similar to fuel injection systems, quite inexpensive and reliable.

Performance figures supplied by System Svedlund AB, for equipment redesigned and tested in prototypes in very recent years, are shown in the table above.

The Saab Model 99L, 1974, is equipped with the smaller Svedlund charcoal system. Its consumption is 20 kg. of charcoal (44 lb.) for 100-110 kilometers (62.14-68.35 miles) at a speed of 90 km/hr (56 mph).

Svedlund made two different systems, for charcoal and wood. Wood

ANALYSIS OF GAS (SVEDLUND)

Gas	Wood Fuel (%)	Charcoal (%)
Carbon monoxide	19	30
Hydrogen	18	7
Methane	1.6	0.5
Oxygen	0.6	0.5
Carbon dioxide	12	1.5
Nitrogen	48.8	60.5

Calorific values   1206 kcal/cu. meter. 1228 kcal/cu. meter

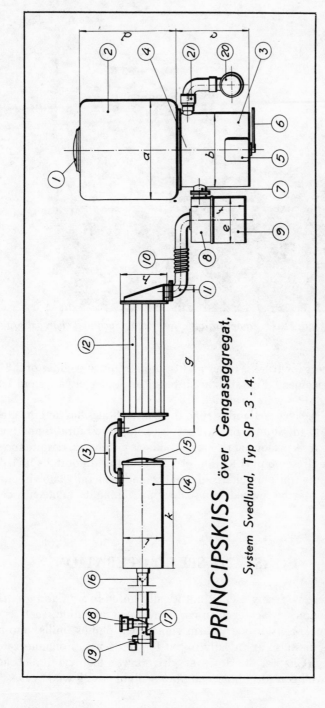

PRINCIPSKISS över Gengasaggregat,

System Svedlund, Typ SP - 3 - 4.

Figure 4-5. Charcoal gas generator, System Svedlund. (1) fuel door, (2) fuel bin, (3) gas producer, (4) inspection door, (5) ash pit door, (6) shaker lever, (7) gas outlet, (8) cyclone purifier, (9) soot bin, (10) expansion tube, (11) gas riser, (12) cooler, (13) cool gas tube, (14) gas purifier, (15) clean-out door, (16) flame arrester, (17) mixing tube, (18) air cleaner, (19) carburetor, (20) electric blower, (21) check valve.

Saab, model 99L, 1974, equipped with the smaller Svedlund charcoal system.

gas producers for trucks, buses and tractors cater to engines of 2.5 to 11 liters displacement. Five sizes are listed. For passenger cars from 1.2 to 5 liters, three sizes were provided.

Because the charcoal industry in the United States has deteriorated to a small, almost invisible regional craft, catering to restaurants and gourmet customers, it seems logical to stress wood gas producers, despite the somewhat more tiresome purification problems of this ubiquitous fuel. When and if the superiority of charcoal fuel becomes generally known, charcoal production may be revived on a scale approaching its nineteenth-century status.

CONSTANT-SPEED OPERATION

The gasogen, with any fuel, is not ideally suited to stop-and-go driving. Its best economy is achieved in constant-speed operation, such as cross-country trucking, stationary farm generators, pumps, mills, and so on. Marine engines also are ideally suited for gasogens, as demonstrated by past history (Chapter 6). However, drivers who log over 10,000 miles a year may well consider the economies available using wood gas.

This wood-powered car, known as the ECAR and developed by ECON Co. of Alexander City, Alabama, has been driven successfully from coast to coast.

COSTS

Rough predictions of the amortization of equipment can be made with a chart (Figure 4-6, page 56), constructed from the following data: gasoline consumption of the car is assumed to be 20 miles per gallon. At 10,000 miles, it will have consumed 500 gallons. At $1.20 per gallon, this comes to $600. Another line shows the increased slope due to a price of $1.60 a gallon.

A wood gas generator is assumed to cost $1,000, on a mass-production basis. Common North American woods, their weights, and relative heat values are given below, all at 12 percent moisture content. A legal cord contains a net wood volume of 90 cubic feet.

HEAT VALUES OF NORTH AMERICAN WOODS

Wood	Weight per Cord	Heat Value, Millions Btu/Cord	10,000-Mile Estimated Consumption, Cords
Birch } Maple }	3,960 lbs.	30	2.50
Oak	4,230	32	2.37
Pine	2,250	18	4.54

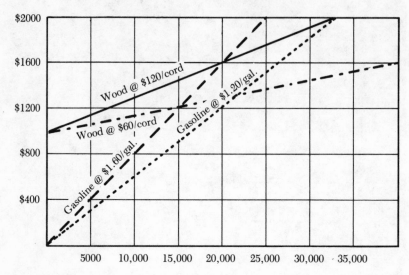

Figure 4-6. Cost of gasogens vs. operating costs.

Consumption of wood in a gasogen is variable between much wider limits than gasoline consumption. Hence the last column is based on an average of one mile per pound of wood. On charcoal, this figure may go up to 1.4 miles per pound, according to data supplied by Svedlund for a model 99L Saab, at 56 miles per hour (90km/hr).

It is apparent from Figure 4-6 that, with current (mid-1980) gasoline prices and high-priced dried cordwood, the gasogen may pay for itself at about 20,000 miles.

CHAPTER 5

Alternative Engines and Fuels

Storms over energy supplies have centered on petroleum and specifically gasoline and its champion consumer, the ordinary internal combustion-engined car. But a slight squall arose around the *external* combustion engine in the mid 1960s. The steam engine is "externally" fired; the fuel either gasoline or fuel oil, is burned at atmospheric pressure outside of a boiler, instead of the compressed atmosphere of the cylinder of a gas engine. The result of the former is that the exhaust lacks most of the pollutants that the Environmental Protection Agency is struggling to eliminate. The hot-air, or Stirling engine, is in the same category.

THE STEAM AUTOMOBILE

Although the steam automobile existed in great numbers and performed creditably in the decade 1900–1910, it used almost the same fuel as its noisy successor, the internal combustion car, i.e., gasoline and kerosene. The "revivals" that have occurred since that era have been statistically tiny and depressing, from an engineering viewpoint, although the popular scientific press has capitalized on the legend of steam cars and the eccentricities of their promoters.

The last revival, actually the final death-twitching of an exhausted subject, was extended by the emissions control movement. To revive a romantic vehicle of the past to meet the clean-air requirements of another era became an exciting crusade for a hundred inventors and thousands of newspaper scribblers. In the momentum of actual prototype construction of cars and buses, it was barely noticed that these steam vehicles were gulping more fuel than their internal combustion competitors — as much as double, in fact. If the Arab oil embargo had not occurred, the twitching might have extended longer.

Despite the flimsy appearance of such cars as this 1902 model, ancient steamers had good performance. This 500-pound model climbed Mt. Washington and Pike's Peak. Models built after 1910 became too complex, too heavy, and too expensive to remain in competition.

Steamers Are Easier to Operate

The legend of the steamer's superiority started in the first decade of the century, with the French Serpollet steamer, the Locomobile, the White, the Stanley, and other lesser brands made in England and the United States. At this stage of development, the internal combustion- or "explosion"-engined-cars suffered several disadvantages. They needed to be cranked. They required priming in cold weather, and more cranking. Gear shifting was a losing gamble. They stalled easily, and their safe management was an art that men would not often entrust to women. The steamer, by contrast, could be fired by lighting a wick. It started, sometime later, by pressure of a small, white-gloved finger on the hand throttle, and it could not be stalled. There were no gears to be shifted, no clutch, and no noise. They were dependable enough to be preferred by many medical doctors.

In the next decade, the self-starter was perfected for gasoline cars, gears became more docile, and spark ignition became more reliable. The steamer lost ground by comparison, and only the Stanley survived the World War I years, limping along on 600 cars a year and the patronage of a diminishing band of supporters, into the middle of the post-war decade, with the Doble flashing briefly across the stage. The once-superior performance of the steamer, its smooth acceleration and indomitable hill-climbing ability, was finally eclipsed by the gasoline engine and automatic transmission.

Steamers Costly

Good steamers were never inexpensive. The large White steamer sold for about $5,000 at a time when the equivalent Rolls-Royce sold for the same amount. In the 1920s, the Doble was priced from $20,000 upwards,

This Stanley Steamer roadster was modernized with Space Age materials.

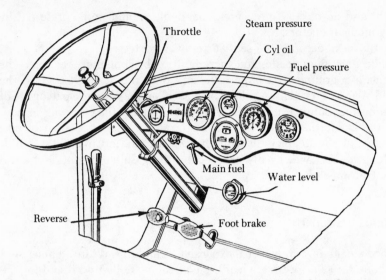

Figure 5-1. The controls of the Stanley Steamer were simple and functional. This is a diagram of the interior of the Model 740 Stanley. The car weighed 3,825 pounds and was priced at $2,750 for the seven-passenger open car, the five-passenger phaeton and the two-passenger roadster. The five-passenger sedan cost $3,585 and the seven-passenger sedan cost $3,985.

depending on the owner's tastes in coachwork. The Doble was a huge car, with one model weighing 6,000 pounds and having a 151-inch wheelbase (see photo, page 61).

To be reasonably efficient, the steam engine must operate at high temperature, and that calls for expensive nickel steels and the difficult machining and assembly of alloys. Only one part of the internal combustion engine — the exhaust pipe — works at a temperature endured by almost the entire boiler and engine of steam plants. Some of the practical problems, such as temperature, lubrication, and routine maintenance, might be solved in production by new materials and techniques, but the basic thermodynamic handicap of steamers cannot be overcome by any breakthrough.

Lear's Experiments

High hopes and fat wallets have set aside the thermodynamic facts, if not overcome them. William Lear, who was happily and profitably building jet airplanes in 1960, decided to try his hand at steam car design.

A charming holdover from the horse carriage days was the whip socket (shown near driver's left hand) on this early Locomobile steam car. In this 1905 photograph, driver is getting water from water spout. A few cups would not take this steam car far, since it had no condenser to circulate and reuse water. (The Bettmann Archive)

This 1925 Doble Steam Car has a 151-inch wheelbase and weighed 6,000 pounds.

With encouragement from ecologically motivated fans, from Edmund Muskie to the kookiest of inventors, he set up shop in April, 1968, to build a prototype steam car and a race car for the 1969 Indianapolis 500-mile race. There was talk of testing a car for the Highway Patrol (state police) in California, and of a fluid to replace water as the power medium, officially called *Learium*, but dubbed *D-Learium* by steam engineers. The man who pioneered the car radio and developed a successful small jet and an automatic pilot was going to extend his magic touch to the legendary steam car.

Several miscalculations and two years later, not much automobile appeared for the $10 million Lear says he spent on development. A "vapor-dyne" car, numbered 23, with a complicated engine and an invisible boiler, toured the auto shows of 1969 behind ropes. Besides a Neanderthal disregard for the laws of heat transfer, the Lear venture foundered on poor judgment—a decision to build a 300 horsepower engine—and on misplaced faith in the ability of employees to design in fields in which they were untrained.

Left: *Steam power isn't confined to automobiles. Wilbur Chapman is shown here with a 300-pound steam outboard engine he designed in 1947. Boiler added another 200 pounds of weight to this power plant.* Right: *The desire for a "modern" steam car drove many mechanics to cannibalize old Stanleys for their engines and boilers. Here is a 1945 special that was assembled by Harry W. McGee (in foreground), using a 1925 Stanley engine.*

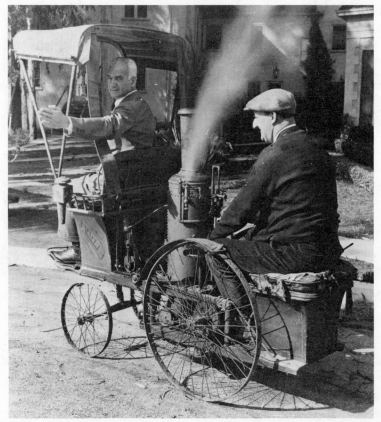

This early American automobile was capable of eight miles an hour. It was first shown at the 1892 World's Fair. In this photo, taken in Los Angeles, the racing driver Ralph de Palma is signalling, while engineer Joseph Wright handles the controls. (The Bettmann Archive)

Buses in California

More significant projects were afoot in 1969. The state of California, where Los Angeles County is the most densely automobile-laden spot on earth (over 3.5 million vehicles at that time), requested proposals for supplying four steam buses for testing. While one-half million dollars was allocated, technologists called for $60 million to develop clean power plants. To make a long story short (every steam story is long), a General Motors bus was running in mid-1972, powered by a Lear steam turbine, with tap-water as the fluid. Although the level of emissions was clearly

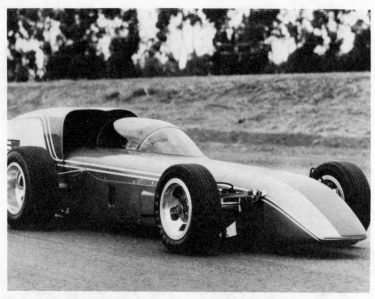

This steam-powered car was designed and built by Skip Hedrich in 1969. (World Wide Photos)

below the 1975 limits, fuel consumption was nearly a pound per horsepower per hour — over double that of the diesel engine it replaced.

When they responded to the pressures to investigate alternative power plants, established automakers in the United States made similar mistakes in transferring gasoline engine horsepower traditions into steam engine requirements. While it is possible for a limited number of affluent customers to support the fuel bills of a 300-horsepower, three-ton gasoline car, a steam plant of this power pushes the physical limits of space available and exceeds, by about 50 percent, the monstrous fuel consumption of the "prestige" and sports cars.

The Saab Project

A different approach was taken by Saab, the Swedish aircraft and car maker. Saab's experimental car, seen by the press early in 1975, had a small engine of fully balanced design and a boiler and control system using latest technology. One of Saab's goals, a lightweight boiler (28 lbs.), is achieved by the use of tubing the size of soda straws.

Although there will be no breakthroughs, the experience Saab gains may give that firm a head start in the event of rapid shifts in fuel economics and synthesis. The steam engine, in addition to its capability of handling great overloads, can use any gaseous or liquid fuel, without regard for octane numbers, emissions control accessories or noise problems. Like many other steam projects, this one was put away in mothballs.

Summary

Measured by the amount of publicity it generated, the revival of the steam engine as a substitute for the internal combustion engine in automobiles and buses seemed as important as the electric vehicle. Research on modern steam cars started with the pollution crisis but ended with the fuel crisis. The dramatic failure of the steam power plants built by the late William Lear to achieve acceptable fuel efficiency levels marked the end of publicly financed research projects.

Except for the restoration of antique steamers, very little activity in steam cars appears. However, Jay Carter Enterprises of Texas, is developing a stationary, solar-powered electric generator. When the fuel is free,

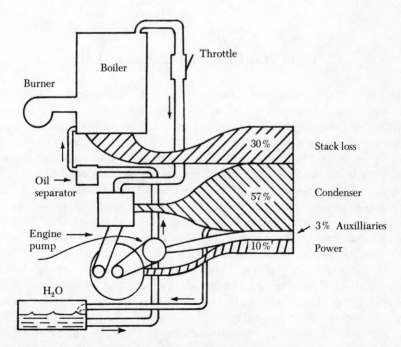

Figure 5-2. This diagram indicates that the steam car's reputation was built on performance, not efficiency. It was the least efficient of all motor cars. Available power was 10 percent of the heat content of the fuel used.

the poor thermal efficiency of the steam engine can be ignored. It is quiet, simple to build, and dependable. The only problem, common to all solar devices, is to store heat for night operation.

Unlike the internal combustion engine, a steam plant of any size can function on any available fuel, from peat and rubbish to hardwood or coal. This versatility alone may bring the once-popular, solid-fueled small steam plant back into use on farms and industries far from sources of electric and oil energy. Members of the two steam enthusiasts' clubs are active in exchanging information, building experimental engines and, in one project, cooperatively producing parts to convert the trusty Volkswagen "beetle" to steam. Membership information may be obtained from:

Steam Automobile Club of America
P.O. Box 529
Pleasant Garden, North Carolina 27313

Steam Power Club, Inc.
1816 Tanglewood Way
Pleasanton, California 94566

THE STIRLING-CYCLE ENGINE

The revival of the Stirling-cycle engine has been unspectacular and uninteresting to all but an engineering minority. The idea of an engine using air instead of steam to drive its piston gained a patent for the Rev. Robert Stirling of Kilmarnock, Scotland, on Nov. 6, 1816. After a modest success as a power plant for butter churns, grinders, and small manufactures until the gas engine came into use, the hot-air engine descended to the status of a toy. It reappeared about 1938 as a silent generator component for military installations. N.V. Philips Co., a Netherlands research and electrical manufacturing concern, improved the efficiency (which was very poor) and reduced the size of the ancient models. General Motors looked into the system for space ships and cars but abandoned it about 1970.

In the mid-1970s, The Ford Motor Co. worked under a Philips license on an unusual car engine in which air was replaced by hydrogen as the working medium. Long ago, it was recognized that air had a poor capacity to transfer heat, compared with steam and other gases, particularly those called "monatomic." But air is cheap and harmless, and the original Stirling plants were safe, if not efficient. Modern engine designers have at their disposal certain sealing techniques that permit the retention of an initial charge of an elusive gas like hydrogen or helium for the life of the engine. The Philips-Ford engine provides speed changes by quick changes in the pressure maintained on a reservoir of the gas. Without this device, heat generated by the fuel burner would be transferred much too slowly through the cylinder walls to accelerate a vehicle in traffic. Conversely, a drop in reservoir pressure makes the gas far less conductive of heat, and causes deceleration.

More efficient than any other external combustion engine, including gas turbines, the Ford-Stirling engine was estimated to cost over $3,500.

Ryder-Ericson hot air engine, built about 1840–50, in the New England Steam Museum, East Greenwich, Rhode Island.

Starting time is long, like the steam system, but there is no problem of freezing and no water and oil separation problems. As in a steam system, the external combustion of fuel leads to very low emissions. It is the only type of engine that is both nonpolluting and capable of considerably higher efficiencies than internal combustion engines now in use. This potential derives from the Stirling's facility for regeneration or recycling of heat from one cylinder to power another, a basic thermodynamic advantage that has been recognized by the Jet Propulsion Laboratory's request for grant funds to develop it, along with turbines, in a 10-year, billion-dollar program.

The basic Stirling engine is remarkably silent. Even with such a high price on this commodity, it is difficult to foresee a commercial automotive use for the system, except for a deluxe limousine. Since the usual fuel for both steam and Stirling engines is petroleum, either kerosene or jet fuel, there is little point in looking to them as oil-savers. However, with a cheap source of non-oil fuel, such as wood gas, methanol, or ethanol, these external combustion systems might become economical, especially where the acoustic environment of a hospital or resort area demanded near-silent power supplies. The Ford-Sterling project has been abandoned.

GAS TURBINE ENGINE

The automotive gas turbine, pioneered by Rover in England, and by Chrysler in the United States, has been plagued by a high count of nitrogen oxides in its exhaust, an inherent feature of combustion under compression. The Environmental Protection Agency awarded Chrysler a contract to seek improvements in this emission problem. Turbines required to propel vehicles at widely varying speeds must necessarily link to driving wheels through high-ratio gearing or electric drives. Hence, the turbine, because of noise, cost of gears, and cost of high-temperature alloys in the engine, is not a likely candidate for replacing the internal combustion engine.

Another impediment to the turbine's use in cars, at present, is its need for petroleum distillate fuel in somewhat larger amounts than luxury vehicles can tolerate. Large marine and stationary turbines may be very efficient, and some of the latter have run on powdered coal. Probably large gas turbines will be coupled with coal or rubbish-fired gasogens in the future, but the small automotive plant seems remote.

ELECTRIC CARS

Compared with the extensive connoisseurship of antique steam cars, there is little popular enthusiasm for the electric, a vehicle associated with little old ladies, shaded urban streets in Boston, Brookline and New York, and, for good reasons, the level landscapes of many Midwestern cities. Its advantages in the first quarter of this century remain much the same today: cleanliness. No exhaust, no dripping fuel or lubricant. No noise. Instant starting, with finger-tip control, and gearless speed changing. Electric cars were, and still are, ridiculously cheap to operate; the nocturnal charge costs a few dimes. They need no tuning, no oil changes, no cooling fluids. In traffic, an electric vehicle (EV), standing still or crawling, uses almost no power, whereas a New York taxicab consumes half a gallon of gasoline and adds to the pollution problem when making a cross-town trip. The EV converts fuel with far greater efficiency and cleanliness than a combustion engine.

There are several hundred thousands of electric vehicles in operation now, all over the world, but most are commercial trucks, vans, golf carts, etc. England is estimated to have 50,000 in service. Many of those in the

This is the front-wheel drive Centennial Electric, *an experimental electric-powered car introduced by* General Electric *in September 1978.*

The GE Centennial Electric has a 24-horsepower DC series traction motor. At left is a 12-volt battery for accessories. Eighteen 6-volt batteries that power vehicle are slung beneath it.

United States are running inside shops, warehouses and industrial plants, and on private property and therefore are not registered or counted in usual statistical summaries. While commercial users may choose electrics for their specific advantages, the motorist commuter or salesman is barred from the choice of electrics by their disadvantages on the public highway. Most electrics have these disadvantages:

1. *Short range, about 20 miles.* This means 20 miles out and return, a more realistic definition than saying 40 miles total.

2. *Poor performance.* Although no electric owner expects jackrabbit starts, he may be embarrassed by toots at the traffic lights and the need to keep near the curb in heavy, faster traffic.

3. *Slow speed.* Although not a necessary limitation, it is a trade-off for range. The electric that can maintain high speed and long range does not exist.

4. *Initial cost.* Electric motors, switch-gear, and especially batteries are all made of fairly expensive metals.

Fossil Fuels Needed

The electric, like the steam car, if it displaced the traditional gasoline car, would solve the air pollution problem pertaining to vehicles, but neither can do anything for the energy problem. Fossil fuels are needed to keep generating plants going, and a significant number of electric vehicles on charge at night, even during "off-peak-load" hours, would erase these slack periods and demand extended plant capacities—exactly what the utilities cannot stand, now or in the near future.

However, when and if these utilities return to the burning of coal, under the refined circumstances of scrubbed and purified exhaust gases, that plentiful domestic fuel can be burned more reasonably and cleanly in huge boilers than can oil in individual vehicles. The electric, although it is unlikely to be as efficient as a directly-fueled heat engine, may thus justify its existence in certain environments on the basis of clean air and silence.

Cars Being Built

There are several electric passenger car projects and a growing production line count indicates that some customers are willing to spend about $3,000 on a specialized second car having severe limitations and gratifying rewards.

One production electric, the *Elcar*, comes to the United States from the Italian shops of Zagato, a well-known coach-builder of sports cars, via Elkhart, Ind. The *Elcar* has 3 to 3½ h.p.; 48-volt, lead-acid batteries; and a maximum speed of 25 mph.

Also in Italy, Fiat is developing a small, ambitious electric with a 50 mph speed and a range of 65 miles. In Japan, where incredibly severe urban pollution encourages government support for electric car research, there is continuing interest in production models.

Toyota has an experimental electric with a range of 60 miles, a top speed of 56 mph, and a braking system that is entirely electric, feeding energy back into the battery. Volkswagen and Daimler-Benz also are jointly involved in fundamental research and plans for electric vans.

The Big Three in the United States have all paid lip service to the concept of the electric second car, but none of them went further than the prototype stages. General Motors knocked itself out of any practical competition by choosing a silver-zinc battery with an astronomical price tag. Ford pinned its faith on a battery using sodium in the molten state—one of the most unstable, explosive metals in the chemistry books. American

Motors likewise cast caution to the winds and built a lithium-nickel battery car.

Making a Better Battery

Obviously the automobile manufacturers the world over are not competent or willing to push the development of a battery that will give both extended range and lighter weight. Battery manufacturers, however, show more interest and ability. The Yardney Electric Corp. of Pawcatuk, Connecticut, already experienced in developing silver batteries for missiles and space craft, is developing a nickel-zinc battery. Gould, Inc., a large manufacturer of conventional batteries and a division of Globe Union, is working on nickel combinations, as they are not too different from lead-acid in construction and service problems.

The truly lightweight batteries are plagued by corrosion, the disadvantages of highly-heated electrolytes, toxicity and high cost. The developers all admit that they are not a sure thing for this decade. The only possible exception is the zinc-chlorine battery, operating at normal temperatures, with materials that are cheap enough to make it competitive with the standard lead-acid battery. The problems yet to be solved are the containment of lethal chlorine gas in case of an accident, and the relatively short life of about 500 recharging cycles.

Despite the re-emergence of a market for passenger electrics, and enlarged research budgets for new battery development, the old lead-acid style will be with us for years to come. It is not too bad to keep the British, who have always managed to make prudent long-term investments in their own high-quality machinery, from using fleets of electric trucks in all cities. The dairy industry alone uses 11,000 in London, with one chain owning 4,500 vehicles. By comparison, the largest fleet now being assembled in the United States, 350 electric jeeps for the Postal Service, seems a chicken-hearted venture.

Congressional Support

To promote the EV, Congress enacted P.L. 94-413 in 1976. It provided funding for long-range research and development programs, as well as for the Near-Term Electric Vehicle (NTEV) program. Many designs for improved EVs have combined mechanical-energy converters, such as the internal combustion engine, with electric propulsion. Hence the law was entitled The Electric and Hybrid Vehicle Research Development and

Demonstration Act. The first step implementing the act was the design of integrated test vehicles (ITV). From several contractors submitting proposals, two were selected to continue to the next phase of actual construction.

The GE/Chrysler Vehicle

One is an advanced experimental electric passenger car with a range in excess of 120 miles, designed at the Jet Propulsion Laboratory in Pasadena, California. General Electric, the prime contractor, and Chrysler, a major subcontractor, developed the four-passenger electric vehicle for the Department of Energy (DOE). The car has been designated ETV-1.

Tests showed that the vehicle, with two passengers aboard, has a range of more than 120 miles at a constant speed of 35 mph and more than 90 miles at a constant 45 mph. During stop and go traffic, the range is about 75 miles. Top speed is more than 65 mph. The ETV-1 is a test car and is not available commercially. However, to assure that it represents an economically attractive car of the future, the car is designed to be suitable for mass production in the mid-1980s at a cost "goal" of about $6,400 (in 1979 dollars).

This vehicle differs from tradition in its double-reduction chain drive to the front wheels, and in its control system. Braking, instead of

This is the new electric test vehicle (ETV-1) developed for the Department of Energy by General Electric and Chrysler.

dissipating heat energy in the brakes, returns it to the batteries (regenerative braking). A hydraulic brake system is linked in, as a backup, and to give proper "feel" to the pedal. Regeneration is effective down to 5 mph, and 55 percent of this energy is returned to the battery. Control of motor speed is through a microprocessor, which is also responsible for charging, braking, reversing, fault protection, and instrument displays. The last includes a "fuel" gauge.

The batteries, of lead-acid type by Globe-Union, are placed in a tunnel on the centerline (fore-and-aft) of the vehicle. Much effort went into achieving an aerodynamically clean body design with a low drag coefficient. In view of the penalties of high speed, particularly in an EV, this effort seems questionable. The streamlining program can be justified, perhaps, as something essential to selling the car.

The Garrett Airesearch Vehicle

The same "fast-back" design is applied to the body of the Garrett car, and lead-acid batteries are used. Similarities end there. Braking energy is fed into a flywheel until a maximum speed of 25,000 rpm is reached. Thereafter, this energy is applied to battery charging. Two identical motor-generators, geared to the flywheel and the driving axle, maintain the proper functions, providing extra power from the flywheel during acceleration and hill-climbing. The second motor maintains flywheel speed at all times except during braking, which uses regenerative energy down to 4 mph. The flywheel efficiency is preserved by housing it in a chamber that is kept under high vacuum by a two-stage electric pump.

The intended effect of this elaborate electro-mechanical system is to keep current flow down to a minimum, thereby extending the range and the life of the batteries. A continuously variable speed changer on the drive shaft, at the rear on this car, works to the same effect. The space taken by this equipment — the flywheel is 23 inches in diameter — seems to eliminate a rear seat or trunk. These two vehicles, when completed, will be subjected to a series of acceptance tests and evaluations of performance, maintenance, reliability, and safety. The DOE has tentative plans for limited production of the prototypes and fleet demonstrations.

According to W.J. Dippold, of the Department of Energy, the best energy efficiency of EVs equipped with the best modern controllers, batteries and motors, can be attained with four-speed manual transmissions. They are better than automatic, or direct drive. Manual transmission lowers battery current demand on starting and improves gradability at speeds of moving traffic. Research should aim to find a transmission with the capability of the four-speed manual type that requires no overt action by the driver.

The clearing house for information on commercially available EVs is The Edison Electric Institute, 1111 19th Street, Washington, D.C. 20036.

Fuel Cells

Battery technology, although it has a history of more than a century, seems to have come to an invisible barrier, forcing scientists to examine the possibilities of converting fuel directly into electrical energy without the need for mechanical and magnetic devices such as boilers, engines and rotating generators. A fuel cell with no moving parts may use a gaseous fuel, hydrogen, hydrazine, or liquid methanol, with oxygen or air as the oxidant. A rather large number of cells is required to boost voltage to useful levels for an automotive motor drive. Although fuel cells are in operation in specialized applications, usually stationary sources of small amounts of electrical current, they are hampered by maintenance problems, and the fact that platinum seems to be the only practical material for electrodes. Hence, the "engine" of a fuel-cell car would cost about $10,000.

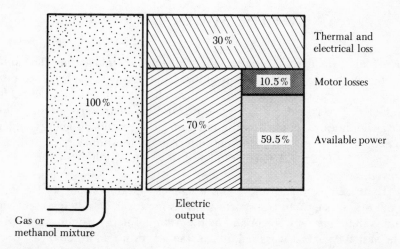

Figure 5-3. A fuel cell vehicle is the most efficient available in terms of output to input. However, astronomical first costs make it a poor practical substitute for the standard product.

There is no noxious exhaust from a fuel cell, the products of the oxidation being carbon dioxide and water vapor. Platinum is being consumed at increasing rates in catalytic converters, and it is unlikely that existing mines could satisfy another demand of any size.

Looking Ahead

The normal internal combustion engine, gasoline and diesel, with devices to control emissions, with basic design improvements such as stratified charge, and with fuel additives such as methanol, is the one sound bet for the general purpose car through the year 2000.

The electric vehicle will grow in numbers, particularly in the form of trucks, as specialized vehicles, and as second cars for local personal transport. Widespread use of methanol blends in the first category would ease present shortages, remove gasoline and oil import problems, and prepare the way for later total replacement of gasoline as a motor fuel.

Steam cars, and those propelled by Stirling engines, fuel cells, and hydrogen will have to wait a long time, either for an improbable technical breakthrough or an equally improbable change in the economic system, wherein waste becomes a virtue.

NEXT: THE AGE OF HYDROGEN

The fuel of the early internal combustion engine was city gas. The automobile became possible when inventors found a way to vaporize, atomize and burn a liquid fuel: gasoline. A gaseous fuel is still very impractical for vehicles. Nevertheless, many scientists and amateurs have become "hydrogenophiles," hoping for the creation of a secondary energy system based on this fuel. The present internal combustion engine will not change over to a hydrogen diet unless a very serious indigestion problem is solved.

Hydrogen is plentiful, it burns cleanly (water vapor is the only exhaust), it is easy to transport, and it is nontoxic. It is being used with high safety scores by NASA and by the Atomic Energy Commission. Over the past 100 years it was used, in a 50 percent dilution, as "coal gas" or city gas. The hazards of this fuel are mainly in the toxicity of the odorless carbon monoxide that makes up a large part of the product. Its disadvantages are in its low heat content (only one-third that of natural gas) and its present cost.

Hydrogen Production

There are at least three ways in which hydrogen can be separated from water, and several ways, in addition to the present internal combustion engine after much modification, that it may be utilized in vehicles. The oldest (1800) and simplest extraction process is *electrolysis*, or the dissociation of water by direct-current electricity. Hydrogen is liberated over the negative pole (cathode), and oxygen over the positive (anode). Although most modern generation of electricity is in the form of alternating current, there are few problems in converting from one to the other. With this system, it is obvious that the cost of hydrogen must always exceed the cost of electricity. However, hydrogen may be stored far more efficiently than electricity and may be transported by many inexpensive means. The theoretical and practical efficiency of the process is high, and the capital costs are within reason — about 45 cents per million Btu of capacity. To be commercially effective, a larger market for oxygen, produced simultaneously and in half the quantity as hydrogen, must be found.

Sites for Power Plants

It may soon be necessary to consider the Arctic and Antarctic zones as the only suitable sites for fossil-fueled and nuclear primary power plants because of present problems of thermal marine pollution as well as particulate atmospheric releases. Also, maximum thermal efficiency is obtained with the lowest possible "sink" temperatures available in unlimited icy oceans. Since these plants would not need to be operated at maximum outputs the year-round (a normal state of affairs), their excess, or peak-load capacity, could be absorbed by hydrogen production with tankers returning the products to populated zones in pressurized containers.

Another method, nonelectrical, for the production of hydrogen has been investigated by a European research team. This is a closed, four-stage *thermo-chemical process* in which heat and water are the only inputs. By eliminating the necessity for electrical power, great savings would be possible. The potential is exciting, as is the prospect for the fuel cell's conversion of gas energy to electrical power without rotating machinery. Both are elegant, philosophically, and elusive, practically.

The third method of hydrogen separation is probably most attractive to solar heat enthusiasts and biologically oriented thinkers. Blue-green algae and other plants are dispersed in cells receiving solar radiation. In the *photosynthetic process*, oxygen is removed from the water, and hydrogen released. Under laboratory conditions, a reasonable efficiency is achieved.

The earth receives 0.8 calorie per square centimeter per minute. The photosynthetic cell yields 500 kilogram calories per square meter per day. Based on this, a commercial hydrogen plant (500 tons/day capacity) would need an area of 14,000 acres, or 22 square miles for solar collectors.

Effort is Slow

With that forward look on record, it is sobering to observe the excruciatingly slow pace of past efforts to capture more energy with less effort. In recent decades, windmills, tidal power plants and thermal seawater engines, using the difference in temperature between tropical surface water and adjacent deeps, have been investigated and tested. Systems depending on weather, wind, and sun must be backed up by conventional emergency systems, and intermittently active plants, (tidal and solar) must have storage facilities. Perhaps the advice of a Committee of Common Market countries should be kept before us in future deliberations on energy sources: Watch the growing role for coal about the year 2000. An estimated 87 percent of the world's petroleum reserves, and 73 percent of natural gas will be depleted by that time, but only 2 percent of coal reserves. Coal can be converted to gas and to liquid fuels.

A very quick appraisal of the huge spaces required and the depreciation on enormous capital outlays expected for these solar and natural systems leads, inescapably, to the conclusions reached by studious foresters years ago: Trees are extremely efficient converters of solar energy to carbon fuels. Recently, it appears that they can be further converted to liquid methanol in portable plants.

A Poor Fuel

Hydrogenophiles generally ignore the fact that hydrogen is an abominable fuel for the present high-compression automobile engine. It causes knocking, poor power output, and has a flammability range in air that makes a small leak an enormous hazard.

	Hydrogen	Butane
Limits of flammability in air, by volume	4%–74%	2%–8.6%

But when and if safe and cheap nuclear power becomes a reality, hydrogen may have a chance as a motor fuel, either in a fuel-cell car or in some new form of combustion engine that may be fed hydrogen from storage in the form of liquid or solid metal hydrides, for example.

Some progress in the latter category may come out of research in metal-air batteries, which are regenerated by hydrogen. Cadmium plates and a platinum catalyst raise the initial cost and skeptical eyebrows, but operating costs might be as little as a half a cent a mile. This compares with fleet operation costs for gasoline vehicles (with gasoline at $1.10 per gallon) of about 12 cents per mile, and electric battery truck costs of 6 cents per mile, including battery amortization. If maintenance costs are included in the mileage costs, the gasoline vehicles trail the electric battery cars significantly, and the hydrogen plant is a totally unknown quantity.

Hydrogen Engines

While hydrogen fuel poses difficult problems for conversion of present gasoline engine designs, the trend to higher petroleum prices, along with mounting costs to meet tight emissions standards, has launched research on four hydrogen engines at Oklahoma State University. Forced injection of hydrogen and delayed ignition are claimed to avoid the usual pre-ignition knocks. To provide hydrogen storage in an automobile equal to a 20-gallon gasoline tank, with presently available techniques, would require a 500-lb. tank for magnesium hydride. This compound must be heated to produce gaseous hydrogen.

There is a recent report on progress in hydrogen preparation from solar sources, from Dupont chemists H.S. Jarrett and Arthur Sleight. They have coated a cathode with rhodate, a rare earth oxide, and when a cell is exposed to sunlight, it not only decomposes water but generates a small electric current. The idea will stay in the laboratory stage until cheaper components are found.

WHAT ABOUT WATER?

What does one say to the proposal that one's car will run on water? Here is a so-called "crackpot" proposal that needs more serious attention than does the revival of steamers.

Although not a fuel, water has returned periodically as an additive to the diet of the gasoline engine. The author's family sedan of the 1920s was equipped with an accessory water jet on the intake manifold, a piece of hose, and a gallon tin can. The gentle spray of vapor — the can was refilled every 20 to 30 miles — quieted the knock of the high compression engine, and seemed to improve performance and economy. But there are other ways to achieve the same results: Rather than using the spray intake, the water (18 percent) has been mixed with gasoline (82 percent) and emulsified.

Test results showed lowered combustion temperatures and resulting low NO_x emission. The University of Oklahoma has joined in a program of testing this idea with the U. S. Postal Service in Norman, Okla. Part of the fleet there was adjusted for the experiment, supposedly on a continuing project.

It is annoying for large consumers of gasoline to calculate how much they might have saved over the years had they replaced 18 percent of their fuel with tap water. My family's estimated waste for a 50-year stint of driving without water comes to about $3,000. Such are the gaps between technology and education.

THE GOOD OLD PISTON ENGINE

When a former cabinet officer urged a gathering of Detroit manufacturing executives to "re-invent the auto," and indicated his belief that the era of the present internal combustion engine was coming to a close, politely suppressed horse-laughs resounded through corporate chambers for weeks thereafter. Professing this belief was a colossal blunder, even for a member of an administration not celebrated for technical and diplomatic proficiency.

There is little wrong with compact car design that could not be remedied by an improved fuel, to reduce pollution and increase performance and reliability. More emphasis on economy, safety and comfort, at the expense of speed, styling, and size, would also be an improvement. With the 55 miles per hour speed limit and the average trip of the American motorist about two miles, it is possible to revise our needs for high-power "gas guzzlers" downward, say from 200 to 100 horsepower or less. City and suburban trips with an adult at the wheel seldom require more than about 20 horsepower at the axle. On freeways and turnpikes, however, it is reassuring, when passing traffic on a hill, to have about 400 percent reserve power. Hence large cars operate at peak efficiency for only a few minutes of every hour on the road, and at a cost of as much as $1,000 annually in wasted fuel.

One solution to the above problem is an experimental Swedish Saab water-injected car. This is based on the stock model 900 sedan, with its normal 2-liter engine producing 118 horsepower. An exhaust-driven supercharger, without increasing fuel consumption, boosts the horsepower to 145. A further gain is achieved, when maximum acceleration is called for, by the injection of water spray into the intake valves. This prevents preignition at higher blower pressure, and raises horsepower to 170, still without a fuel consumption penalty. The 44 percent power increase, although it involves some expensive accessories, could mean considerable savings in fuel over a year.

PERPETUAL MOTION

The search for new sources of abundant cheap energy has accelerated work by inventors whose enthusiasm exceeds their knowledge of physics. Sometimes the demonstrations of motors running by permanent magnets, gravity, or the spin of the Earth are good enough to reach the popular press. Patents are issued, models built, and investments are made. Thus far, the only energy created is that involved in moving money from the safety of deposit boxes or mattresses to the pockets of inventors, attorneys, and modelmakers.

CHAPTER 6

Past Uses
of Synthetic Fuels

"Why didn't someone warn us?"

Since the energy crisis and recent gasoline shortages, we have heard this many times. The truth is, of course, that we *were* warned — many times.

In the past similar warnings have been issued — and acted on — in other nations. A look into their histories shows that while these nations may have reacted to the warnings in different ways, nearly all made some positive move to assure themselves of a continuing supply of fuel for transportation.

In the period between world wars, realistic European nations knew that World War I hadn't been the "war to end all wars," and they prepared for yet another. Some nations concluded that a major war would cut them off from supplies of oil. All knew that the successful prosecution of a major war would require an ample alcohol supply both for fuel and for the manufacture of high explosives.

FUEL BLENDING

The pattern in most of Europe was the same — more and more measures to force the motorist to use a substitute fuel blended with gasoline. By 1937 this program, which subsidized the alcohol industry, cost European governments $234 million. That year 18 percent of European motor fuel was produced from wood, coal and agricultural products. Imported gasolines were selling for 9 cents per gallon before taxes, but alcohol in 11 countries averaged 44 cents per gallon.

Alcohol was not the only substitute used. Synthetic liquid fuels from coal and gas, generated on board, were others. Thus, in 1937, European consumption of substitute fuels was:

Fuel	Metric Tons, Thousands
Alcohols	510
Synthetic gasoline, from coal (by Fischer-Tropsch process)	929
Benzene	824
Shale oil	38
Compressed gas & wood gas	234

While the driving public was given little choice in the matter, being forced to purchase the more expensive blends if they desired to continue to drive, it should not be inferred that they were purchasing an inferior fuel, even in the critical 50/50 blend. The blended fuel could provide more power, reduce or eliminate knocking, and produce a cooler and less noxious exhaust. It could eliminate the addition of tetraethyl lead, an advantage not recognized until recently. Its only disadvantage, hardly noticed by those who could afford to own automobiles, was its increased consumption compared with pure gasoline.

The German Experience

In view of the dominant role of Germany at the outbreak of World War II and her lack of a major oil supply of her own, it is not surprising that she made the most significant quantitative steps toward an independent fuel supply.

The processes for manufacture of synthetic fuel from coal that were developed in those pre-war years in Germany are the ones that hold the most promise for those in the United States urging use of coal and municipal wastes for this purpose. They are the Lurgi process, now being revived in Frankfort, and the Pott-Broche, the Bergius-Pier, and the Fischer-Tropsch processes, all developed prior to the Nazi era. When that shadow fell, a dozen plants were created, and by 1944, they were producing over 4 million tons (32 million barrels) of synthetic oil and gas annually from coal.

With so great a need for fueling a war machine, a military dictatorship did not reckon costs only in marks; thus it would tolerate the process inefficiencies that resulted in expensive fuel in terms of coal consumed.

Today, with oil prices rising, Germany is reconsidering a return to her deep coal mines for another go at synthetic fuel production.

While Germany's pre-war production of motor fuels was not impressive when compared with wartime needs, the figures do illustrate the percentage of home-produced fuels, figures that increased geometrically through the war years.

GERMAN DOMESTIC FUEL PRODUCTION, 1932

Benzol (from coal)	200,000 tons	44.6%
Alcohol	80,000 tons	17.7%
Gasoline	170,000 tons	37.7%
	450,000 tons	100%

This total represented only one-third of Germany's vehicle and aircraft fuel consumption for the year, with the remainder furnished by such friendly nations as Rumania.

The first German government decree on fuel called for the addition of 2.5 percent alcohol to all imported and domestic gasoline. This was later elevated to 10 percent. It was enforced with difficulty, due to inadequate supplies of good-quality methanol. Two common fuel blends, Monopolin and Bevalin, called for 25 percent alcohol and 75 percent gasoline; a third, Aral, was composed of 20 percent each of alcohol and benzol, with the remaining 60 percent gasoline. In the days just prior to the outbreak of war, the nation specified that all motor fuels would be blended from 70 percent gasoline, and 10 percent each of ethyl alcohol, methanol and benzol.

But Germany's enemies-to-be were taking action, too.

France

Soon after the cessation of World War I, France moved to a major domestic grain alcohol fuel industry by establishing a 50/50 mix by volume of ethyl alcohol and gasoline, calling it "carburant national," and decreeing that it be used by government departments and Paris buses. The government required the oil companies to buy the alcohol, to blend it, and to suffer any resulting losses.

In 1931 the law was broadened to require all commercial vehicles to use a new blend, called "heavy carburant national," that used 25–35 percent alcohol. Shortly after passage of that law, another was passed that reached into the tanks of private autos. This law called for the addition of

10 percent alcohol to *all* imported gasoline. Although annual national production of ethyl alcohol was less than 200,000 barrels — France had not yet taken to private wheels — the legislation seemed to have achieved its purposes: to reduce losses in foreign exchange to oil-producing countries and to bolster the domestic ethyl alcohol industries for military exigencies. Complaints were frequent about the legislation, and its result, desultory performance with "carburant national." The half-and-half mix made it quite difficult to start one's vehicle on a cold morning. About 25 percent alcohol, as shown by many tests, is the upper limit at which the blended fuel will vaporize satisfactorily in an unmodified carburetor.

England

England's early situation was slightly different. With her global oil companies firmly established, the Imperial confidence was high enough to shun legal measures until Britain entered World War II.

In the mid-1930s, two alcohol fuels were available in England. Cities Service Co. produced "Koolmotor," a blend of 16 percent ethyl alcohol and 84 percent gasoline. Cleveland Petrol & Distillers Co., Ltd., produced "Cleveland Discol," a racing blend composed of about 79 percent ethyl alcohol, 9 percent acetone, and 10 percent gasoline.

Other European Pre-war Measures

Most nations on the Continent lacked the British confidence (and with good reason) and were cutting gasoline consumption years before their roles in the war-to-be were settled.

Beginning in 1931, the minister of finance in Austria had the authority to compel the addition of alcohol to gasoline when the wholesale price of the former was lower than the latter. A top limit of 25 percent alcohol was set, and the law required domestic alcohol. The modest result of this temperate law was the production of 6,786 hectoliters (180,000 American gallons) of alcohol in 1931–32.

In 1932, Yugoslavia made it illegal to drive with any mixture other than 25 percent alcohol and 75 percent gasoline, and Poland compelled fuel dealers to add from 6 to 12 percent alcohol to gasoline.

Czechoslovakia, with a higher degree of industrialization, and a firmer attachment to the private vehicle than her neighbors, began mandatory use of alcohol fuel in 1928 with a blend called "Dynalkol," consisting of 50 percent alcohol, 30 percent gasoline and 20 percent benzol. Annual consumption was about 2½ million U.S. gallons in 1932, when all refined imported fuels were required to be blended with 20–30 percent alcohol, consisting of 95 percent ethyl and 5 percent methyl alcohols. The latter was made by wood distillation.

Secondary goals of the law were to increase alcohol production and potato farming. Seven wood-distillation plants were producing 35,000 hectoliters (924,700 U.S. gallons) of methanol per year.

In 1931, Italy had a law that called for 25 percent ethyl alcohol being mixed with all imported gasoline. Latvia passed a similar law, with purchase of alcohol being through a state alcohol monopoly. Hungary's regulations provided for a 20 percent alcohol mixture with gasoline in a blend called "Motalko." Diesels and farm vehicles were given exemptions from using this.

Scandinavia

Sweden and other Scandinavian countries had limited highway systems and no crude oil production in the 1930s; they thus used little motor fuel compared with today's consumption. Sweden did offer various blends, called "Lattbentyl," of which 170,000 barrels were used in 1931. The Swedish people made alcohol from waste sulphate pulp liquor. The important part that the process could play in pollution control in the American paper industry is now being recognized.

Africa

South Africa was producing blends as early as World War I. During that war the Union of South Africa manufactured and sold "Natolite," a mixture of 60 percent ethyl alcohol and 40 percent ethyl ether, both sugar products. At the end of the war the components changed to 50 percent ethyl alcohol and 50 percent gasoline.

South America

Some South American nations, too, moved toward blends in the between-wars period. Argentina had enough petroleum to forestall such legislation, despite agricultural pressures to add alcohol made from grain, grapes and sugar cane.

Brazil's production of alcohol in 1932 reached about 16 million gallons; and she had no production of gasoline. Despite this imbalance, little effort was made to force regulations upon the affluent motoring minorities.

In Pernambuco, two interesting blends were sold. One was a blend of 70 percent alcohol and 30 percent ether, and the other, sold in Rio de Janeiro and Nictheroy and called "Gasalco," was a blend of 82 percent alcohol and 18 percent gasoline.

Ether was needed, instead of an integral heater, to enable the engine to start without pre-heating either the intake air or the fuel. Alcohol, when used as a pure unblended fuel, does not readily vaporize in the standard carburetor, but when heated adequately, it becomes an ideal, cool-burning fuel. It is doubtful if the rich Brazilians, in their European sports cars and limousines, took advantage of the high-octane rating afforded by this racing fuel by increasing the compression ratios and advancing the spark timing.

Uruguay, Chile and Salvador, with state monopolies for fuel, had alcohol-additive laws similar to those in Europe.

ALCOHOL FUEL IN THE UNITED STATES

While nations in Europe and South America achieved varied success with alcohol-gasoline blends in that between-wars era, the achievements in the United States were minor, localized and short-lived.

Alcohol as a lamp fuel had a brief history in this country. Through the first third of the nineteenth century, whale oil was the preferred fuel for domestic lighting. With the increasing productivity of farms, grain was plentiful enough to provide alcohol as a lamp fuel, and the odorous animal and fish oils were replaced. The change emphasized three important qualities of alcohol fuel: It is odorless, nontoxic when burning, and the spilled fuel, if ignited, can be extinguished with water.

Then petroleum was discovered in Pennsylvania in 1859, and kerosene, because of its very low price, displaced alcohol as a lamp fuel, despite the odor of the new oil.

Dual Fuel System, 1900–1920

The early history of use of alcohol in the automobile in this country indicates that several early American automobiles were equipped with a carburetor heater and dual piping for alcohol and gasoline, thus providing complete interchangeability in fuels. The sectional drawing (Figure 6-1) from a maintenance text of 1911 shows a flexible fuel system, as well as illustrating the options in auxiliary equipment available to customers. Alcohol fuel equipment and instructions appear frequently in the texts,

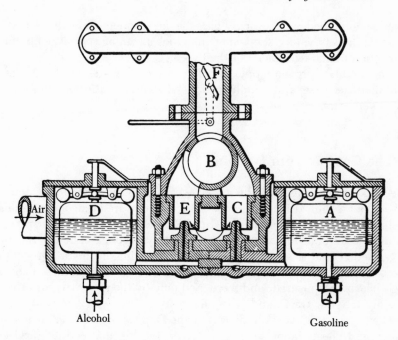

Figure 6-1. A double float-feed carburetor for alcohol and gasoline. The engine was started on gasoline from chamber A, *and when the engine was hot, the rotary valve* B *was turned by the driver to use alcohol fuel from chamber* D, *drawn through jet* E. F *is throttle valve.*

leading the reader to conclude that the fuel must have been accepted as equal to gasoline in most markets.

In these first two decades of the car, one of the principal advantages of alcohol fuel, its compatibility with gasoline, was not utilized. Gasoline was used to start the car, and pure ethyl alcohol was used to run it, when the induction system became warm enough to vaporize the alcohol.

Now it appears more prudent to blend alcohol with gasoline, extending supplies of the latter and gaining, together, the merits of each fuel used separately. At some future date, when oil runs out, we may return to the pure alcohol vehicles of the past.

Tried After War

The first post-war wave of alcohol fuel blending came in 1922–23 when large stocks of World War I alcohol, made for munitions production, became surplus at low prices in the Baltimore area. Post-war motoring demands had pushed up the price of gasoline, and production facilities lagged behind. One of the largest oil companies took advantage of this temporary disruption to market a blend of the two fuels.

The results were unhappy. Alcohol absorbed water in the dealers' tanks, and led to separation, plugged fuel lines and erratic performance in customers' cars.

The Conflict with Gasoline

Although the causes of this fiasco were obviously avoidable, factors other than technical incompetence altered the future of alcohol. Among them was progress with gasoline. After 1923, new oil finds, the general expansion of the cracking process increasing the yield of gasoline and the use of tetraethyl lead as an anti-knock agent (replacing alcohol at one-tenth the cost) all lowered the price of gasoline. For about a decade alcohol fuels nearly disappeared, and were used only by the race drivers in their cars on state fair tracks and at Indianapolis.

Then came 1933 and the depths of the Depression with its farm surpluses and cheap labor to put otherwise idle distilleries into service. Idealistic and ecologically alert farmer groups encouraged government support of farm-produced fuel projects. They were successful in several states. First the Iowa legislature proposed a law requiring the blending of alcohol with gasoline fuel. Then Nebraska and South Dakota removed state fuel taxes on the alcohol portion of blends.

But there were more failures than successes. Twenty bills in Congress and 31 in state legislatures failed to pass. Despite the low prices of grain, alcohol cost more than gasoline, and the ecologists, who wanted to spread organic fertilizer made from spent mash over the fields, began to give up. Politicians who had advocated the farm subsidies inherent in alcohol fuel production turned their attention to more direct methods of cash subsidy and support.

Private Efforts

There were efforts outside of government circles. In 1936 the Chemical Foundation, a Delaware corporation then thriving on impounded German patents and an enthusiastic agrarian reform philosophy, set up a distillery at Atchison, Kansas, to make "power alcohol" from farm surpluses. Atchison Agrol bought molasses, corn at 28 cents a bushel, and sold blended fuel for 25 cents a gallon. The plant closed after 2½ years of unprofitable operation.

Many reasons for the failure were advanced: The equipment was worn out at the start; the price Agrol paid for gasoline was too high and the federal Surplus Commodities Corporation failed to cooperate by selling grain at a price below the market.

A more ambitious experiment was tried by Herman Willkie, brother of the 1940 presidential candidate Wendell Willkie, and an officer of the Joseph E. Seagram and Sons Corporation. Willkie initiated research projects in his company and in the Indiana Farm Bureau. He tested tractors using alcohol blends, analyzed and improved processes of fermentation and distillation of alcohol fuel, and published a book, *Food for Thought*.

Willkie claimed to have research farms where sweet potato production reached 500 bushels per acre, releasing 500 gallons of alcohol. Some of the nitrogen removed from the soil by this intensive farming was returned by spreading the by-product mash as fertilizer.

Two requirements of Willkie's manifesto could be applied now to the agenda of the advocates of methanol utilization: (1) "A simple and portable continuous-process operation for conversion of the crops into alcohol" [substitute "organic wastes" for crops, today], and (2) "Internal combustion engines that will operate efficiently on [pure] alcohol."

Portable Distillery

For the first requirement Seagram drew up plans for a five-car railroad distillery, to operate continuously, with grain-to-alcohol conversion time of about eight hours. Other crops could be taken in, with ease. Willkie reported experimental use of wood wastes to produce fuel at 20 cents a gallon.

The project for the second requirement materialized modestly in a few prototype tractors by International Harvester, modified to high compression, for use in the Philippines. Another demonstration project, a Chrysler-engined tractor, was modified and tested for Seagram at the University of Kentucky.

Willkie's trainees at Seagram, from many of the "have-not" nations, particularly India and South American countries, were encouraged to return to their native lands with the technologies of advanced farming, natural (by-product) fertilizers, and farm-fueled tractors. However, famines and the economic distresses of these countries, resulting directly from fertilizer shortages and energy pricing, all related to petroleum imports, rang a sad note of defeat to Willkie's high aspirations.

ORIGINS OF THE GASOGEN

The idea of harnessing blast furnace gas seems to have occurred to M. Lebon about 1800. He had invented illuminating gas and suggested that it might have been used a century earlier by those precursors of the internal combustion engine (Papin, Huygens, etc.) who dared use gunpowder as an engine fuel.

Lebon's idea lay unused until the development of Jean J. E. Lenoir's engine in 1860. But it was on other minds, not the least facile of which was Beau de Rochas's. He described the "four-stroke" cycle in 1862. Had the French been more aggressive, and had this scientist's name been shorter, the cycle might not have been called the Otto cycle.

Several experiments contributed to the development of the modern "gazogene" in the decades prior to the important techniques of liquid fuel carburetting by Otto, Benz and Maybach, and the subsequent successful launching of the automobile on its errant course. Alfred Wilson and Emerson Dowson developed furnaces in which air and steam were injected into the combustion zone. The gas, known to the French as "gaz pauvre" (poor), had a low heat value because of the dilution of the principal combustible, carbon monoxide, with carbon dioxide and nitrogen. However poor it may have been, it was the first gas — a real gas, not the liquid we casually call gas — to feed the very first internal combustion engines, made by Jean Lenoir in Paris. It was also the first fuel to be manufactured from raw materials on board the moving vehicle, in a "gazogene portatif," at a later but uncelebrated date.

Lenoir's Drawing, 1860

The automobile of Jean Lenoir has been the subject of extensive historical speculation. It may not have existed in 1859 except in the minds of Lenoir and his overzealous press agent and fund-raiser, Gautier, and in Lenoir's patent application of that year. However, a drawing published in *Le Monde Illustre* in 1860, showing a high wagon, a one-cylinder horizontal engine, and a riveted gas pressure tank, established certain facts about the automotive art of the time — or the lack of it.

Although historians point out that the car could not have run because it lacked a clutch and other modern essentials, these questionable details are submerged by the well-documented fact that compressed coal gas was the fuel of the hour. The company formed by Lenoir's competitor, Hugon, was named "La Compagnie du Gaz Portatif" (Portable Gas Company).

This gas was produced in a retort similar to those that continued to operate in some communities, where coal was economical, well into the twentieth century.

An Engine in Action, 1861

In 1861, Lenoir had the opportunity to install one of his engines in a launch being built by another inventor in the gas business, Lasslo Chandor. Gas for illumination was becoming popular, but if one's house were beyond the gas mains, what could be done to keep up with the urban fashions? Chandor gas was distilled from naphtha and turpentine, and, presumably it was dispensed as a liquid, to be re-evaporated at the gas lamp. Although it was a clumsy way to provide gaseous fuel to a mobile engine, the hot tube probably used in this instance antedated the liquid carburetor, and might have made the automobile possible. Chandor and Lenoir certainly share honors for designing the first internal combustion motor boat. There is evidence that it navigated the Seine, but no data on performance are available.

Again traversing the thin ice of historical priorities, it may be claimed that Chandor gas was the first instance of the blending of a fossil fuel (naphtha) with a renewable fuel (methyl spirits and resins from wood, or turpentine).

There was no novelty in the production of city gas in the works of Paris and other capitals. A retort with a trap door and funnel at the top was the basic furnace; the result was piped into scrubbers, filters, tar separators, and so on.

The Portable "Gazogene," 1914

In the form of the portable "gazogene," the gas producer made its reappearance only in 1914. The prototypes for the portable equipment were found in the well-tested stationary gas producers of small size that made their appearance at the expositions of 1889 and 1900 in Paris. Instead of producing gas for storage in a tank, called a "gasometer," these more modern units were "direct induction." Professor Arbos, of Barcelona, was credited with this idea in a paper of 1862. It will be seen that the output of the generator will be in proportion, more or less, to the engine's speed, since the engine provides suction to maintain the fire in the generator.

The gasogen eliminated the bulky gasometer, with its water seals, and it produced gas, under slight vacuum, as was required by the engine (see Figure 6-2). Prior to this discovery, the injected or blown gas producers

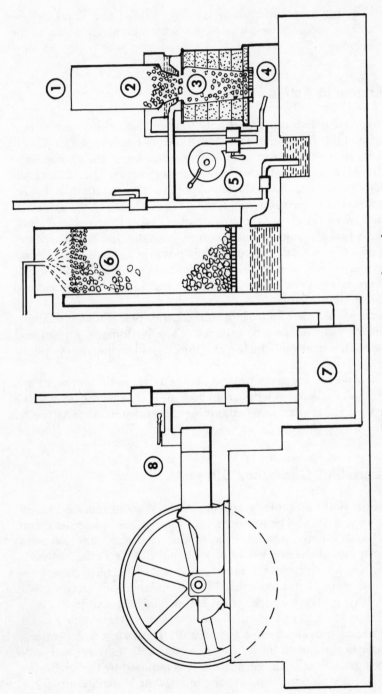

Figure 6-2. Where city gas was not available to run small industrial power plants, this self-contained suction gas producer and internal combustion engine were used. The fuel was coal. (1) filler cap, (2) coal magazine, (3) combustion, (4) ash pit, (5) starting fan, (6) scrubber, (7) dry filter, (8) throttle.

94

used petroleum, which may seem strange to us. However, oil was easily vaporized and metered into the furnace at a constant rate. Remember that the carburetor had not been perfected for use of volatile liquid fuel in an engine.

THE FIRST PORTABLE GAS GENERATOR

"Poor gas" gained a foothold as an internal combustion engine fuel, in the typical stationary installation, and the solid forms of carbon, particularly charcoal and coke, were found to be more reactive than oil in a small retort, such as might be carried on a horseless carriage at a later date. The first truly portable vehicular gas generator to be installed in the United States was in a boat, the launch, "Gloria," in 1908. Fifty-four feet overall, she was fitted with a four-cylinder engine and a coal-fired gas generator.

Fishermen and other commercial boat owners, converting their craft over from steam to the new gasoline and oil engines, felt the pinch of high fuel prices about 1910. This increase was reflected by the production in Boston of a line of gas producers for marine use. The Nelson Blower and Furnace Company, with engineer A. L. Galusha, equipped new and old boats up to 175 horsepower with money-saving gas generators until oil and engine prices dropped after World War I.

Economic Units

Although the Galusha gas systems appear bulky and heavy by modern standards of marine power plants, the owners judged them by comparison with the much heavier steam boilers and engines which were replaced, and their economy was evident. A construction launch of 50 horsepower, for example, cost $1.80 per hour to operate on gasoline in 1914 (at 12 cents per gallon). If the system were converted to pea anthracite coal (at $8.25 per ton), hourly cost was about 20 cents or one-ninth the gasoline cost. Now, with the cost of the two fuels elevated five times, the ratio remains about the same.

THE FIRST AUTOMOTIVE GAS GENERATOR

By a trick of fate, the romantic city of Casablanca became the site of the launching of that most unexciting, cumbersome, automotive anachronism of the first half of the twentieth century — the self-motivated gas generator. Under the sponsorship of the Automobile Club of Morocco, five trucks and tractors took part in a series of tests for farm vehicles during the war of 1914–1918. From this historic occasion until 1936, there are records of many efforts by the French Ministry of War and by other bureaus of research and invention to popularize the "gazogene" in a country that had no liquid fuel of its own, but those practical people, obsessed with tourism, deluxe motor cars and racing, took a dim view of the cumbersome and temperamental "ersatz" gasoline machine.

The rallies went on for another decade in France. In 1925 there were 17 vehicles, French and Belgian. The road test, over a route of 2,100 kilometers, was followed by bench tests of six hours for each vehicle. The period was the peak of "gazogene" activity, with 25 types on the market.

Despite the stimulation of rallies sponsored by the French army and the agricultural bureaus, there were only about 2,000 "gazogene"-equipped trucks in use in 1938 — almost on the eve of World War II. There were very few innovations in French equipment in this final period, and, of course, none during the Occupation.

Problems During World War II

War in Europe brought severe belt-tightening, if not total curtailment, of all gasoline use by noncombatants. In 1940, gasoline prices were up to $1.03 per gallon in Italy, then a neutral. Alcohol was blended with petroleum fuel to extend it, and gas generators appeared in large numbers in every country in Europe. Germany's restrictions were the tightest: no civilian driving, no new automobiles, no uncontrolled use of gasoline. However, a synthetic fuel industry was very rapidly developed, using coal as the raw material. Even far-away New Zealand felt the pinch, and gasoline was rationed at 8 to 12 gallons monthly to civilians.

Britain

In Britain, where the domestic gas industry first flourished in the nineteenth century, the use of this commodity as a substitute motor fuel was the obvious answer for a coal-run country without oil. Gigantic gas bags, at atmospheric pressure, appeared atop buses and cars. They contained the equivalent of a half-gallon of gasoline, and they needed filling at frequent intervals. Trucks and buses often towed a trailer with a gas generator to retain adequate space for payload.

France

Although French reactions before the war were apathetic to the whole-sale conversion of commercial and military vehicles to solid-fueled gas generators, their production of them in 1941 escalated remarkably. About 20 percent of the normal motor fuel needs were met by substitute fuels in that year. The Automobile Club of France announced that there were 60,000 charcoal-burning cars in operation and that 40,000 more were in production. Forestry reserves could support 300,000 vehicles thus equipped, it was estimated. Where youth camps replaced military service, charcoal production was the chief occupation, and in 1942, a total of 36,000 tons of charcoal per month was being produced.

Ethyl alcohol production for motor fuel blends was 10,000 tons that year, and this was expanded later by increased planting of alcohol-producing crops, notably sugar cane and beets.

A Maintenance System — Brazil

Brazil, with no important oil reserves of its own, anticipated its wartime problems by government encouragement of charcoal production and government-sponsored design of "gasogenios." In Sao Paulo, a commission of 60 was established. It trained 1200 servicemen for maintenance of the equipment. This wisdom paid off and made it possible to operate transportation into the interior, where petroleum fuel would not be a factor for years, or decades.

Scandinavia

Scandinavia, with its forests and its insignificant fossil fuel reserves, developed some excellent gas generators. The Svedlund System, made in Sweden, included a small boiler that injected steam with the air. The water thus introduced into the reaction zone was dissociated in hydrogen and oxygen, the latter combining with the incandescent fuel as CO. However, the variables in the process are too many in a small unit and the actual function of the steam was to reduce the temperature of the fire bed so that fusion and clinkers were avoided. Later models of the Svedlund System omitted the steam injection.

METHANE PRODUCTION

Sewage gas, containing enough methane to power an engine, has been seldom used because of its lengthy gestation in the works, about 30 days. One plant in Erfurt, Germany, compressed it into metal cylinders to 200 atmospheres (about 3,000 psi), enabling cars to make between 150 and 250 kilometers (91 to 155 miles) on a filling. Coal, charcoal and wood chips, because of the convenience of self-service and low price, remained the mainstay of the gas producers' industry in most countries of Europe and in South America.

CHAPTER 7

The Future of Cars and Fuels

Auto manufacturers have spent millions making experimental gas turbine cars, steam cars and exotic engines. Government agencies and urban organizations have spent or will spend a billion dollars to attempt to clean up the air, to undo the damage of years of neglect. Car owners are paying dearly for catalytic converters that are suspect of emitting an acid pollution of their own, and they are paying a price for gasoline that will only go up. Billions (of decreasingly valuable dollars, to be sure) are going to the Middle East and Africa for oil and for platinum. To maintain the flow of this single potent energy source, we have even whispered of military methods.

Despite all the research, expenditure, and risk, the manufacturers agree that no really radical change in the basic internal combustion-engined car will be made before 2000 A.D. There are economical and practical ways in which present fuel supplies may be extended. There are ways to phase in unfamiliar fuels without disrupting the economy — at least not to any greater extent than it is at present. We have seen that alcohol fuels and wood gas generators could help extend petroleum supplies and reduce the volume of imports. Another way to further this aim is through taxation.

THE EFFECTS
OF GOVERNMENT TAXES

Although most Americans are only recently aware of government efforts to control foreign exchange deficits, particularly with the oil-exporting countries, Europeans have had far more painful reasons for complaint since the early 1930s. Then, a sampling of vehicle taxes was:

Country/State	Average private car tax per hp*	Actual tax on a 50 hp Rolls-Royce
Austria	$11.80	$590
England	11.80	590
Italy	11.65	580
New York and Pennsylvania	1.65	83

*This horsepower was computed from a formula established in the age of steam.

At that time, vehicle taxation in England brought in a revenue of $138,500,000; and fuel tax receipts were $176,120,000 for a total of $314,620,000. The average total taxation per vehicle was $145 per year.

Those levels of taxation, in addition to reducing fuel imports, produced a hidden beneficial result: They induced manufacturers of cars to reduce the displacement (size) of engines and to develop the high-speed, high-compression sports car engine and chassis that became famous in England and Europe long before Detroit made feeble efforts to gain a fraction of the imported car market.

PATHS OF ACTION –
ALCOHOL AND POWER PLANTS

While high taxes and raised prices on petroleum fuels have reduced gasoline sales, we still need to pay more attention to synthetic fuels from renewable sources. Two parallel paths of constructive action appear in the search for a synthetic automotive fuel policy. The first is the development of alcohol production, stepped up both for blending and for future 100 percent alcohol fueled engines. Second is continued development of nonpetroleum power plants of the large, stationary kind now generating electricity, with concern for by-product fuels for vehicles. The primary energy may be nuclear, solar, tidal, wind, water, or geothermal. The part-time, or off-peak-load by-product might be electric current for vehicle batteries, metal salts for conversion in the vehicle to hydrogen gas, or other safely and easily handled liquids or gases on which conventional vehicles will run.

Another way to run a vehicle on stored energy, rather than burning fuel in the car, is to energize a built-in flywheel at stations equipped with electric or other motors. This has been accomplished on a Swiss bus line, and might be adapted to other short-haul services, such as suburban shopping and school vehicles. Another way to use flywheel energy is to couple the braking system of a vehicle to its power source. This is achieved in the Garrett electric vehicles.

Cellulose to Glucose

Although ethyl alcohol is man's second-oldest conversion product — sour milk was probably first — a way of making it from hitherto-unyielding cellulose, from wood, from paper, and from other renewable materials was developed in 1974. An enzyme was created by Dr. Mary Mandels and John Kostick of the U.S. Army Natick Laboratories that would break down cellulose to glucose in a few hours. Glucose is a sugar that converts, with yeast, to alcohol. Biological conversions of this kind are vastly more efficient in the heat they require, than the hydrogenation of coal or the synthesis of methanol.

A most promising source of cellulose for the long run (municipal rubbish is only about 50 percent cellulose) is manure, available from cattle feed lots in huge quantities. It is nearly all cellulose.

Ethanol and "HydroFuel II"

The product of this new biological conversion process is mainly ethanol, with the impurities that render it unfit for drinking but make it an almost ideal additive for gasoline motor fuels. The superior mixing qualities at low temperatures and the higher heat content of ethanol over methanol are being utilized in tests and promotion of a proprietary product called "HydroFuel II," developed by United International Research, Inc.

This fuel consists of about 30 percent crude alcohol (including fusel oil, *t*-butanol, and up to 10 percent water) with about 70 percent gasoline, stabilized with an unstated amount of "Hydrelate," a synthetic fluid compatible with the others. While automobile test results are impressive, the vertical integration of biomass sources in a dispersed geographical pattern necessary to market the product is not likely for the near future.

Methanol for Diesels

The qualities of methanol fuel that make it ideal for spark-ignition engines — anti-knock, heat absorption, slow ignition — prevent its use in a diesel engine without modifications. However, a research project at MAN

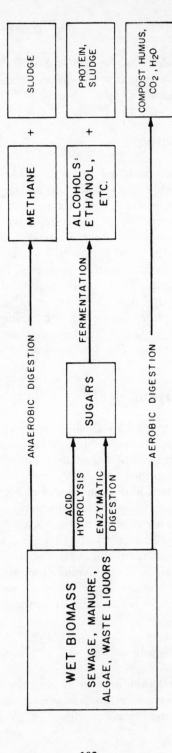

Figure 7-1. Conversion routes for wet biomass, such as sewage, to fuels.

(Maschinenfabrik Augsburg-Nurnberg AG), aimed at reducing dependence on imported fuels, has also improved certain diesel operating characteristics. An ignition accelerator was added to the fuel, and preheated methanol was injected separately from oil fuel, later in the stroke. The experimenters reported "complete lack of soot in the exhaust gases" and an increase in torque without overloading. However, there are still problems with emissions of CO and HC.

Volkswagen Tests

Volkswagen has tested a gas turbine engine using methanol and found it to run with lower nitrogen oxide emissions than are produced on jet fuel. NO_x emissions over the legal limits were among the problems arising with Chrysler's extensive testing of turbine cars although this was not a factor in the postponement of manufacturing. Rotary engines, those without reciprocating pistons, have potentials for fuel-saving and reduced emissions regardless of fuel used.

Continuous Internal Combustion

One system of great promise — theoretically, at least — investigated long ago for steam turbines is called *continuous internal combustion*. In a contemporary hypothetical case, a gas generator using solid or liquid fuel — either coal or methanol — is mated with a rotary engine or turbine. In such a closed system, gas quality can be controlled to refined limits; heat exchange is nearly perfect; there is no pre-ignition problem; all fuels are usable; there is minimal exhaust noise; and production costs should be low.

Volkswagen scientists have mentioned, in connection with this "dream-car" system, a combustion catalyst that would dissociate methanol, providing hydrogen to the heat inputs.

Gas generators of "on-board" types deserve the revival of interest shown in a recent report from Germany. Their advantages, particularly with internal combustion engines, are clearly evident: liquid fuels can be "cracked" before entering the engine, hence eliminating nearly all emissions problems and all "skipping" or misfiring with lean mixtures. As these advantages become more important, along with use of coal and wood fuels as economic conditions indicate, the gasogen may come into the research lab for modernization and automation of its diverse functions.

THE INTERNAL COMBUSTION ENGINE

Although we frequently heap curses upon it in times of failure, the internal combustion engine is a remarkably versatile performer, exerting itself imperceptibly to leap from sluggish urban traffic to mountain climbing at 60 miles an hour. Its tastes for various fuels are unbiased. It runs fairly well on natural gas, coal gas or sewage gas, on its weekday diet of gasoline or diesel oil, and, on a spree at the race track, drinks high-proof vodka or methanol (ethyl and methyl alcohol).

Such flexibility is surprising, in view of the development of the engine as a constant-speed power source to run such machines as saws, forge blowers, butter churns and hammer mills, and to be fed solely on illuminating gas (made from coal) through a pipe.

The gas engine was the brain child of the French mechanic Lenoir, who built 100 of them for Parisian customers. It used the two-stroke cycle, and had such an extravagant appetite for coal gas that most of them were soon converted to steam. Otto took up the idea, using a four-stroke cycle, still with gas fuel. Daimler added high-speed rotation; Benz developed electric ignition to replace the flaming wicks and red-hot igniters of Otto. Maybach contributed the float-feed carburetor, and a fully mobile power plant was ready for the horseless carriage. In two decades of feverish invention, Lenoir's clanking toy was metamorphosed.

The last major change in the automobile came around 1905, when the numbers of steam cars on the road were exceeded by cars with internal combustion engines. A glance at the patent records of recent years indicates that we may see more radical changes in the coming decade than occurred during the past 75 years. The rotary engine was the one exception, and it is in embarrassed retirement. The reason for this plateau was, of course, the low cost of fuel in the United States. Miles per gallon, which had averaged around 30 in 1905, dropped to about 10 for the behemoths of 1973. There was no incentive to produce efficient cars in the United States.

However, there was notable progress abroad, in countries where the price of fuel was double that of ours.

Fuel System Advances

The stratified charge idea, patented by Nicklaus Otto, was hauled from its moth balls by Honda engineers. The "CVCC" (Controlled Vortex Combustion Chamber) made it possible to run a clean engine without anti-pollution accessories. Its regular carburetor takes in a very lean mixture, and then a smaller carburetor and separate inlet valve surround the spark plug with a rich mix that is sure to ignite.

Several small European cars are equipped with fuel injection. This device not only improves reliability and performance but saves fuel. Each measured shot enters the cylinder at the exact moment it is needed, reducing smoking and over-rich idling and coasting pollution. The venturi carburetor, past its 100th birthday, is sure to be replaced soon. There are several competitors seeking to fill the job.

A Princeton professor of mechanical engineering and aerospace science, Enoch Durban, believes that the basic piston engine is an elegant and adequate machine. But the controls, such as fuel metering, ignition timing and valve action, are primitive. He has invented a device to correct the sloppiness of fuel injection. When a carbureted car is accelerated, the fuel-to-air ratio is usually too lean. The carburetor lags several seconds, until it senses the shortage and responds. Upon deceleration the mixture is too rich for another period. Durban compares a carburetor to a stopped clock: it is "right" only for a moment twice each day.

On Durban's Toyota test car, the flow of air is measured at each cylinder. In the absence of details, we must guess, from hints in the description, that the air is ionized, and a counter at the inlet adds up the number of molecules that have passed and triggers the injection of a proportional amount of fuel. The accuracy of the device is so good that a Pinto-sized car might get 40 miles per gallon and run without pollution controls. Volkswagen has arranged to start a test program with Durban's system.

Another hopeful inventor in the race to make the conventional carburetor obsolete is William Henry Beekhuis, a Californian whose four-cylinder compact car is fitted with an "acoustic" injector system. An aluminum box, containing four plungers, replaces the usual inlet manifold. While the word *acoustic* is somewhat misleading, a metaphor is useful in describing the working principle. The plungers "hear" the position of the intake valves, react to the high velocity of the air, and squirt the proper amount of fuel into the cylinder. The car used by Beekhuis was running on methanol fuel.

Devices that improve the economy and performance of gasoline engines obviously adapt to alcohol fuels. The difficulty of distributing alcohol vapor to several cylinders in equal amounts without condensation, has been mentioned elsewhere in this book. By eliminating the need for an injection pump with a mechanical timing drive from the engine, the Beekhuis design seems to have undercut the potential cost of efficient injection by a considerable margin.

Improvements in the spark ignition system of the gasoline engine are coming into use slowly. Electronic surveillance of the engine's condition, or computerization, is available on many cars. However, the gains in fuel conservation will be limited until new hardware, like Durban's injectors, is available to respond to the data-collecting black boxes. An engine accessory that has been long overlooked is the valve train. For a hundred years, poppet valves have been leaking, burning, sticking, and wasting quantities of unused fuel. At high engine speeds, they float beyond the camshaft, dwell, bounce, and waste more fuel. But they are cheap to produce, and therefore have become "the standard of the industry." Long ago, we had rotary valves that required no reversals of energy to drive; they were quiet, positive and efficient, but slightly difficult to lubricate.

As the need for improved automobile mileage increases, it's likely that rotary valves and other earlier innovations will get a new look from some of the automotive engineers in Detroit.

WHAT MORE CAN BE DONE?

There are other steps that could be taken to increase the efficiency of American cars.

Reduced weight. Unnecessary weight in a car lowers performance. While this is known by all automotive engineers, many of the cars of the early twentieth century were lighter in pounds per horsepower than present cars. The cost, in energy, of recycling junked and rusted steel shells is a hidden horror of programmed obsolescence originating in the styling parlors of Detroit.

Improved styling. Sometimes styling is said to reduce air resistance, thereby improving mileage. Has it?

The Rumpler sedan of 1921 had an average drag coefficient of 0.5, meaning half the resistance of a block of the car's dimensions. To achieve reasonable mileage at high speeds, a coefficient of 0.25 to 0.30 is urged. Certain U.S. dragsters and European sports cars have gone down to less than 0.2, and the standard French Citroen sedan has had 0.3 for years. American cars have shown little progress from the Rumpler sedan of more than 50 years ago. Present body styles of the American cars have an average drag coefficient of 0.47. The result of our neglect of genuine streamlining may be a few hundred gallons of gas per year per car extra, and extra billions of gallons wasted by the nation.

THE API

Working quietly behind the scenes in Washington, lobbyists of the American Petroleum Institute try to see to it that its members are in no way threatened by legislation that might limit their profits or their freedom in perpetuating a nearly complete monopoly of world oil supplies. Their only failure, the inevitable termination of the depletion allowance, came about because most taxpayers could understand the inequity of this loophole.

The API is dominated by the "Big Seven" companies — Exxon, Mobil, Standard Oil of California, Standard Oil of Indiana, Texaco, Shell, and Gulf. The institute's budget is supported by the dues of more than 265 corporate members and 7,000 individuals. The institute's role in influencing Congress is significant.

Technical and statistical services of the API are maintained by a staff of about 330 employees, through publications, a speakers' bureau, and 11 lobbyists who supply congressmen with facts and figures not available from government sources.

The smaller oil companies, those not "integrated" in drilling, refining, distributing and selling their products, are organized in the Independent Petroleum Association of America.

The major oil companies will, according to Walter J. Levy, an economist, "continue to be the most important technical and marketing force in most of the world for a long time. Their technical competence and logistical services can only be replaced at a great risk to those who eliminate them." Serious though it sounds, this threat may be weakened by factors that outreach the great powers of the international companies.

The Challenge of an Alcohol Industry

The strength of these companies depends, in varying degrees, on producing, transporting, refining, and marketing oil products. A domestic and regionally dispersed alcohol industry, however, needs no prospecting, drilling, or pumping equipment. Raw materials — coal, wood, or rubbish — can dictate a close proximity to a synthesis plant, and no tanker fleets are necessary to bring the product to market. Two of the four critical activities of oil companies are thus obliterated. The synthesis plant for methanol is no more costly than the equivalent oil refinery, and some of the country's 200,000 filling stations are already selling alcohol blends, mostly Gasohol. The use of alcohol will expand naturally as the qualities of the fuel become widely evident.

And a decentralized ethyl alcohol industry is being revitalized. Older distilleries shut down because they were uneconomical are coming back on line to produce alcohol fuel instead of alcoholic beverages. Instead of using high-quality grains as a feed stock, they use spoiled grains or fruits, cheese whey or other agricultural products. Flavor and purity are of no interest to the palates of our cars. Moreover, ethanol plants may use coal for their process heat, and even solar heat for distillation.

CHAPTER 8

What's a Driver to Do?

The American motorist stands at a curb beside a favorite possession, the automobile, and doesn't hide his concern. Gasoline prices have soared since those ominous days of the 1973 gasoline crisis. And while he still doesn't understand *why* there was a gasoline shortage, this motorist knows that the forces that locked the tanks and cut bulk gasoline deliveries then can do it again. Shortages reoccurred in 1979.

He asks the questions that all of us who drive cars frame in our minds: "What can I do?" And, "Can I avoid any future gasoline crisis — and still drive?" The answers aren't simple. But there is something he can do.

An Electric Car or Gasogen?

If his commuting distance is relatively short, 20 or so miles a day, and he has an extremely small family that prefers a short Sunday drive, perhaps an electric car would be a good buy for him. It will be turtle-slow rather than jackrabbit-jump at the lights, but he can ignore the gas stations, opened or closed.

If he is of a mechanical bent, he should consider the possibility of a gasogen. Although difficult to design, construct, and maintain, the gasogen will make him independent of petroleum fuels.

Conversion Kit?

Or, if he believes bottled gas is here to stay, there are conversion kits on the market today so that he gets his car fuel supply in a bulk tank. There's no major drawback to this system, unless it's the possibility of a gas as well as a gasoline shortage.

109

Alcohol Fuels

But our motorist, there on the curb, asks about the alcohol fuels, ethanol and methanol.

He's heard they're dependable fuels, could be relatively cheap fuels, and could be made from any varieties of "trash," of which we have a good supply in this country. "If it's so good," he asks, "why can't I get methanol or ethanol, and use these fuels as a blend in my present engine, or have a new engine that will burn straight alcohol?"

The answer is that ethanol, mostly in the form of the blend, Gasohol, is available at an ever-increasing number of gasoline stations. But methanol is another story, even though many experts regard it as the more promising, renewable fuel.

A LONG WAIT FOR METHANOL

Methanol in this country for car fuel? Don't park your car in a metered zone while you wait for it, or you will push a lot of dimes into that meter. The major oil companies are less than enthusiastic about any move toward methanol, and those in government seem peculiarly content to go along with the oil companies, even while loudly proclaiming their concern over energy shortages to any who are still listening.

Opposition from Established Companies

The oil companies attempt to quiet any irreverent consumer curiosity about the alternative fuel, and they are anxious to localize the controversy to highly technical and often irrelevent theoretical matters, aimed above the average motorist's comprehension.

One shot was recently heard beyond the laboratory. A fleet test for alcohol blends was cancelled by MIT at its Energy Laboratory under circumstances that suggested pressure by corporations (Exxon and Ford) that support the research facility. Both MIT and the corporations denied that influence had invaded the high temples of university research.

This doesn't answer why the use of methanol seems to be opposed by American oil companies, and is ignored by United States government agencies set up for energy research and management. One would at first think that the oil companies would welcome a fuel that could be made from coal (much of which in this country is owned by the oil companies). This use, it would seem, would improve the performance of the gasoline with which it was blended, and eventually, would take the place of gasoline.

Since the top decisions of the oil companies are made in the privacy of their board rooms, one can only speculate on the reasons that have determined their policy against methanol.

Here are a few of the possibilities:

Competition. One is that manufacture of methanol and ethanol is relatively simple compared to the manufacture of gasoline. Furthermore the raw materials for methanol and ethanol, whether coal (sunshine stored for millions of years) or biomass (sunshine stored this year in cornstalks and hay or this century in wood) are widely distributed and accessible to many more groups than oil has been. Thus chemical companies, the government or new corporations could go into competition in supplying energy. The oil companies see themselves as the sole major purveyors of energy to consumers and have no interest in this technological opening to competition.

Capital. Another is that old invested capital is a universal barrier to the acceptance of new technology. When one builds a new plant, one convinces the bankers that this invested capital will be paying itself off in ten or twenty years, and one convinces one's stockholders that the profits will continue in perpetuity. The one threat to the argument is invention, which promises a cheaper or better product and makes obsolete previous plants. In this light, an oil company director would argue that profits are rising nicely now and let's get present equipment paid for by oil before we allow new technology and invention to make obsolete present plants and equipment. Since alcohols burn cleaner than gasoline there could be a hue and cry by the populace, demanding conversion to alcohol as soon as possible, thus saving the oil for its valuable chemicals, if the truth about methanol became widely known and accepted.

Exploration. Finally, the appearance of alcohols on the market could interfere with exploration for new oil. As long as there is no alternative to oil to keep us warm and moving in our cars, we are likely to pay any price for oil exploration, the greatest single expense of petroleum and gas production now. Of the wells completed in established fields in 1974, 37 percent were dry. In new fields, called "wildcats," 90 percent were abandoned, and only 2 percent proved to have more than a million barrels of oil.

As the odds against finding new oil go up, so does the cost of exploration and drilling. Adding to the instability of such high-risk ventures are the anxieties and ecological costs of drilling in the Santa Barbara channel and on the Grand Banks, and the hazards of supertankers on the shores of Alaska and the world's oceans.

CONVERSION TO METHANOL

Our concerned motorist has yet another question. If oil companies refuse to move toward conversion to methanol, why don't chemical companies take this step? Why aren't new corporations formed, solely to develop and put on the market this new fuel?

The answer today is a matter of simple economics. Despite the increase in gasoline prices and the probability that ultimately methanol will be cheaper, today it is more expensive than gasoline. Thus there is no incentive for other companies to set up the huge manufacturing and distribution system for a product that could not be priced competitively at this time. It should be noted that the experiments in West Germany and Sweden are being underwritten in part by those governments.

The Cost

Before we become too pessimistic about replacing oil with synthetic fuels, let us assess the cost of doing this and how much it will hurt our pocketbooks. The problem is clear: We must build plants sufficient to manufacture about 100 billion gallons of liquid fuel per year, which is the rate of present gasoline consumption. And we must do this within the next 30 years, the period during which the remaining oil resources of this country are expected to be used up.

A number of recent estimates have put the cost of making synthetic fuel at about $1 billion for a plant capable of making one billion gallons, and we will have to build 100 equivalents of this—$100 billion—over the next 30 years. What an enormous sum of money, we say. And yet this is the cost that we have incurred in building our interstate highway system, and it hasn't been excessively painful. We presently pay a federal gasoline tax of 4 cents per gallon for those highways, and since we use 100 billion gallons of gasoline a year, this collects $4 billion a year. A similar tax could pay for the creation of the synthetic fuel plants that will continue to make our highway investment useful and would collect $120 billion over the 30-year period during which our oil is expected to run out.

Other Ways

We might raise this money in other ways. For example, as gasoline becomes more expensive, we could continue to sell synthetic fuel at a related price, with increasing profits. The profit from the early synthetic fuel plants could be used to build subsequent plants. Eventually the price of synthetic fuel would fall below that of natural gasoline and at that time no more gasoline would be produced. The remaining oil could be saved for valuable chemical production.

Like complacent political candidates, the oil companies are the incumbents, with a going concern, to state the situation mildly. At this time, there is no profit for them in a change of fuels.

This stand should not surprise us, since we have seen it before. In 1949 technological data were accumulated on the performance of alcohol fuels during the wartime years. The reports that came out of these data were strikingly different. In countries having ample oil supplies, the reports on performance were negative. But those countries without such resources reported good results from alcohol fuels.

Since the testing methods on which these reports were based left something to be desired, "one can scarcely avoid the conclusion that the results arrived at are those best suited to the political or economic aims of the country concerned, or of the industry which sponsored the research."*

A POSSIBLE DREAM:
ENERGY SELF-SUFFICIENCY

In the period 1973 to 1978, the pessimists of the petroleum shortage were in command. There seemed to be no way to encourage voluntary conservation. Oil experts said that alternative motor fuels could not be produced in adequate quantities before the year 2000. They were equally positive that these fuels could never compete with gasoline. There was apathy and depression everywhere, except in the research offices of a few government agencies, and of a few oil companies.

We had almost given up hope of a sound national energy policy. Investors were not interested in taking the risks of unfamiliar processes in new plants to supply synthetic fuels. Then, with state-sponsored pro-

---

*S.J.W. Pleeth, "Alcohol — A Fuel for Internal Combustion Engines" (London: Chapman Hall, 1949).

grams to produce alcohol fuel, the grass roots revolution was launched. The grain-farming states reinvented Gasohol, a blend that was popular in the 1940s as Agrol. States with more natural gas than grain promptly set up programs to demonstrate the virtues of methanol as a blending extender and as a total replacement for gasoline.

If the quantity of substitute fuels produced remains low, the psychological effects of this production have been enormous. There are still critics in high places who, like the little old lady seeing her first airplane overhead, exclaim that "them things will never fly." Even the staid *Wall Street Journal*, a confirmed sounding board for petroleum industry spokesmen, turned the corner on January 31, 1980, with an article admitting that Gasohol "shows enough promise as a way to stretch dwindling petroleum supplies that U.S. scientists are working hard to make it more cheaply using less energy."

Gasohol is being sold in more than 40 states. Ethanol plants are being expanded. Every few weeks the price of gasoline climbs a little closer to the price of ethanol. In a single week there were news announcements of the planning of three large methanol plants in North America. The huge W.R. Grace & Company embarked on a joint venture to exploit coal fields in Colorado for a synthesis plant. The goal is 5,000 tons of methanol per day. A large Boston engineering firm has picked a waterfront site near Fall River, Massachusetts, for a coal-based gasification and liquid fuel plant. A joint venture of three companies plans a methanol plant near Kitimat, British Columbia, using natural gas to produce 1,250 short tons of methanol per day. We will need dozens more of these coal- and gas-fired methanol plants, hundreds more large and small ethanol plants, before we can think of becoming independent of imported oil. And we will need more diesels, more small cars, and all the gas-saving ideas in this book, and more; we will need restraints, voluntary and economic, for conservation. We will need to expand all mass transit facilities. But energy self-sufficiency is not an impossible dream.

One Hundred Percent Alcohol Cars

Many racing cars, sports cars, and family sedans are running now in different parts of the world on what the critics like to call "exotic" alcohol fuels. These cars demonstrate that pure alcohol works in ordinary spark-ignition engines without pollution or mechanical difficulties. They also suggest how simple it would be to produce engines designed specifically for alcohol.

Drivers who have their own source of ethyl alcohol or have access to a supply of methanol will find the conversion of a small car to alcohol no more difficult than "souping up" a stock car for track racing.

"MX-100" SYNTHETIC FUELS TEST CAR

In January 1978, the California Legislature authorized a state-sponsored research team to purchase two Ford Pinto station wagons, one to be converted for alcohol, the other to serve as a comparative standard, using gasoline. The alcohol fuel chosen was 90 percent methanol and varying amounts of ethanol, isobutanol and n-propoanol. Methanol fuel containing these ingredients can be made more inexpensively than pure methanol.

Three Engine Changes

Three major changes to the engine were made:

- The compression ratio (CR) was increased from 9:1 to 14:1.
- Peak engine torque was shifted (by camshaft replacement).
- The fuel system (carburetor) was replaced by an injection system to optimize methanol use.

Roberta Nichols, project engineer and consultant on California Synthetic Fuels Test car program. The Mx-100 Pinto behind.

The efficiency of engines increases as the compression ratio increases, but, with gasoline, so does knocking. Methanol, with an octane rating of 106, is knock-free at the 14:1 compression ratio. An increase of 7 percent efficiency was expected from this ratio increase alone. For structural reasons, the CR increase was not obtainable by shaving the cylinder head, as is customary. Hence new pistons were designed and made to order by a firm specializing in racing engine work.

The shift in peak torque was made to improve performance at lower engine speeds (to match more closely the stall curve of the torque converter of the automatic transmission). This did not impair top-speed power, and with the increase in CR, gave an estimated torque increase at the flywheel of 18 percent. These two changes, expensive on a one-shot basis and not essential to make the car operate, would add nothing to the cost of an engine put into production. In fact, the displacement, size, and weight of a 2.3-liter engine might be reduced, and still give performance equal to the original gasoline engine.

Changes made to the fuel system do not fall in the same category. A simply-modified carburetor might have sufficed for the demonstration. However, the project designers wisely tackled the alcohol fuel induction problem from scratch. Methanol's high latent heat of evaporation, so advantageous from the point of view of safety, and good combustion when it is once inside the cylinder, has always made it difficult to vaporize and distribute properly in an intake manifold exposed to different environments. Because several light and inexpensive European cars have solved similar problems by fitting injectors in lieu of carburetors, injection was adopted for the Mx-100. Nozzles at the intake valves open at 250 pounds per square inch, giving fine atomization* of the fuel. The injection pump is a discontinued model of a Bosch six-cylinder unit, with a new camshaft for four-cylinder use.

The calibration of the injector involved road testing the modified engine with the original carburetor, in place, to monitor fuel consumption, pressures, speed, and load. Then, on a test bench, the injector was set to operate in conformance with road conditions. To adapt the Pinto carburetor for methanol fuel, the jets listed below were used.

Circuit	Primary	Secondary
Main	0.074"	0.066"
Idle	0.070"	0.041"
Air bleed	blocked off	0.066"

The addition of the injector, driven by a timing belt, necessitated the relocation of the alternator from the right to left side. The exhaust manifold was replaced by separate pipes leading to an exhaust collector. The separate pipes permitted improved exhaust flow and a means for individual instrument measurements.

---

*To atomize is to reduce to minute particles or to a fine spray.

Fuel injector, re-worked unit from an old sports car, supplies methanol to Mx-100 Pinto. It is driven by a timing belt.

Problems and Solutions

This program was completed in two months, with an appropriation of less than $50,000. Along the way, problems were discovered but solutions soon found:

1. The lining of the standard fuel tank, a lead-tin mixture plated on the steel sheet from which tanks are fabricated, was decomposed by methanol. The fuel filters and pump became plugged with the sludge. When the lining was all dissolved, the trouble stopped.

2. The injection pump was noisy. Methanol has no lubrication quality. One percent diesel fuel added to the fuel tank eliminated this problem.

3. Starting was no problem at temperatures of 40° to 110° F., at altitudes from sea level to 8,000 feet. But at less than 40° F. cranking time exceeded 20 seconds, and was considered intolerable. The propane bottle, fitted with solenoid valve and a push button on the instrument panel, admits gas to the intake manifold. Starting time, with propane admitted, is less than with gasoline.

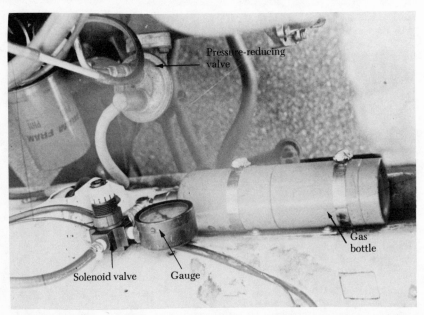

Cold-weather starting device for car using pure methanol: solenoid valve (left) is activated by button on dash; gauge shows level of liquid petroleum fuel; pressure-reducer valve leads to air cleaner and carburetor intake.

4. Fuel consumption was initially high. Pessimism guided the selection of too-large jets for the mixture controller, and 9 miles per gallon was recorded. Smaller jets, although they did not spoil the high performance, reduced consumption to 14 to 15 miles per gallon. This is equivalent to 28 to 30 miles per gallon of gasoline, on an energy-per-volume basis. (The gasoline-burning model recorded about 21 miles per gallon.)

5. Emissions. No external control devices are fitted to the Mx-100. The following data were taken with the on-board exhaust gas analyzer:

Driving Mode	CO, percent	HC, Parts per million
acceleration	0.1–0.4	0–200
cruise	0.2	150–500
50–60 mph cruise	0–0.01	50

The carbon monoxide (CO) figures compare with the satisfactory emissions of a diesel. Hydrocarbons (HC), on the other hand, are difficult to control with methanol and must be reduced. Meanwhile, a catalytic converter would make the HC emission acceptable. NO_x emissions from methanol-fueled engines are so low, (proven in previous tests) that they were not measured in this series.

DESIGN CHANGES FOR FUTURE
METHANOL-POWERED CARS

In future production-model, methanol-powered cars, the material for fuel tanks will be changed. Epoxy resin coating is an obvious solution to inside corrosion. The propane bottle for cold starts will be replaced by an electric heater to vaporize a small quantity of methanol. A block heater is already available for very cold climates. The internal aluminum parts of the fuel pump showed some signs of corrosion, and would be anodized.*

Gasoline and Methanol Compared

I was able to test-drive the Mx-100 within the confines of a small state park. The slow-speed performance was notably superior to any comparable four-cylinder car fueled with gasoline. There was none of the hesitation and roughness that is characteristic of cars with lean-burning carburetors during rapid acceleration.

---

*To subject a metal to electrolytic action as the anode of a cell in order to coat with a protective film.

PINTO COMPARISONS

	Gasoline	Methanol
Vehicle costs	same	same
Maintenance	higher	lower
Btu per gallon	120,000	60,000
Compression ratio	8.5:1	14:1
Octane rating	90	105
Btu per mile	5,455	3,529
Miles per gallon	22	17
Cost per gallon*	72¢	40¢
Cost per mile*	.033	.024
Annual fuel cost*	$495	$360
0–60 mph	18 seconds	16 seconds

The table of comparisons and the substance of this appendix have been reported by Dr. Roberta Nichols, consultant to the State of California on the project. She presented her findings at the Third International Symposium on Alcohol Fuels Technology, Asilomar, California, in May 1979. Dr. Nichols, the recipient of an advanced degree in combustion engineering, and an amateur race driver, was well-qualified for her part in this program, conducted by many individuals and the Mechanical Engineering Department of the University of Santa Clara.

*Estimates made in 1978. While fuel prices are going up, the proportional advantages of methanol are retained. The gasoline-methanol comparisons point to the need to shift one's criterion of fuel economy from miles-per-gallon to miles-per-dollar data.

The same advantageous data would apply to a 100 percent ethanol-fueled car. Because of California's adequate supplies of natural gas, the source of its methanol, the latter was chosen for this state-sponsored project.

APPENDIX B

The Report of the Alcohol Fuels
Policy Review

The Alcohol Fuels Policy Review was established in July, 1978 in the U.S
Department of Energy, to explore and report on the potential of alcohol
fuels. Certain policy recommendations have already been endorsed by the
President. The report was issued in June, 1979. It consists of 119 pages,
more than two-thirds of which are appendices, made up of factual
material, in texts, tables, and graphs. It is available from:

National Technical Information Service (NTIS)
U.S. Department of Commerce
5285 Port Royal Road
Springfield, Virginia 22161

Printed copy: $7.25
Microfiche: $3.00

SUMMARY OF FINDINGS

Alcohol fuels, ethanol and methanol, can contribute to United States
energy resources. Ethanol is commercially available (as Gasohol) and will
be the only alternative fuel that can be on the market in quantity before
1985. At this time, ethanol might displace as many as 40,000 barrels of petroleum per day. If plant construction were planned promptly, methanol
might be produced from coal beginning in the mid- to late-1980s.

No one energy source can solve our national problems. Alcohol fuels
can be but a minor solution in the short term, and hardly a total solution
later. But blended with gasoline, they can extend United States oil supplies and improve the octane rating of unleaded gasoline. Gasohol can be
used in existing vehicles, without modification.

121

Methanol enhances octane similarly to ethanol. But present production cars using concentrated methanol must be inspected for damage to plastic or rubber seals, and to parts of lead, zinc, and magnesium. Both types of alcohol permit the use of higher compression engines, the compression ratio increasing proportionally with the concentration of alcohol with gasoline.

Alcohol's octane-boosting properties make it attractive as an additive because high octane gasoline is in short supply, and other boosters, lead and MMT (methylcyclopentadienyl manganese tricarbonyl) are banned or restricted because they contaminate catalytic converters of new cars. Two other boosters, TBA (tertiary butyl alcohol) and MBTE (methyl tertiary butyl ether) are permitted, but are made largely from petroleum. Ethanol has been permitted by the EPA as an additive.

The production and use of Gasohol, for most of the 1980s, will be limited by the capacity of facilities to convert raw materials into ethanol. The supplies of materials are more than adequate to meet the needs through the mid-1980s. Ethanol-fuel production (about 4,000 barrels per day) is expected to increase to 20,000 barrels per day in 1982. This will place Gasohol at 3 billion gallons per year, or 3 percent of present gasoline consumption. This increase may come largely from unused distillery capacity and expansion of others. If the present federal motor fuel excise tax exemption for fuel containing biomass alcohol is extended, as proposed by the president, investors will be encouraged to build new facilities, and ethanol production might reach 500 to 600 million gallons per year. This would displace up to 40,000 barrels per day of petroleum.

By 1985, costs of ethanol made with improving technology should be reduced, while gasoline prices will increase. Methanol from biomass and coal will become more important in the long term, as stationary turbine fuel and for vehicles designed for its special advantages. Methanol plants must be large—20,000 to 50,000 barrels per day is considered minimal. If and when methanol becomes the principal vehicle fuel, ethanol will remain as a blending component. Both alcohols are being used experimentally as extenders for diesel fuel.

Federal Incentives

Among the federal incentives for alcohol production are these:

1. National Energy Act motor fuel excise tax exemption on gasoline/alcohol blends, worth 4 cents per gallon of blend, and 40 cents per gallon or $16.80 per barrel of alcohol in 10 percent blends.

2. Eligibility of alcohol fuels for DOE entitlements, worth about $1.00 per barrel of ethanol or 2 to 3 cents per gallon.

3. Loan guarantees for alcohol pilot projects; U.S. Department of Agriculture.

4. Investment tax credit, 20 percent, in Title III of the Energy Tax Act of 1978.

5. DOE's research and development funding programs; alcohol fuels budget up to $24.9 million in fiscal year 1980.

6. The Economic Regulatory Administration of the DOE has adopted pricing regulations to encourage Gasohol production by allowing retailers to pass through the cost of ethanol. Other encouragements are planned on both federal and state levels.

Demand For Gasohol

While the method of launching a Gasohol industry was being discussed in 1978, the market developed spontaneously. Retail outlets number over 2,000 in 40 states as of 1980. Several states have exempted Gasohol from local taxes. Iowa seems to have led other states with a 6½ cents per gallon exemption, and sales of 5.6 million gallons in March 1979.

Two major auto manufacturers have included the use of Gasohol in their warranty policies. Three are involved in alcohol fuels research, and one is working on a straight alcohol engine.

ALCOHOL FUELS POLICY ISSUES

The possibility of a conflict between the needs for food and for fuel is often raised. However, the feedstocks for alcohol fuels, during the 1980s, will probably remain wastes from agriculture, distressed products, and by-products. The by-products of alcohol from grains, DDG (distillers' dried grains) are important as protein feeds for animals, and might become human foods in time. Cellulosic materials, not useful as food, and coal and peat, can be processed, in the future, to alcohol fuel. Cellulose, converted by enzyme action to sugar, may provide ethanol. Solid fossil fuels, through gasification and catalytic conversion, can provide methanol.

At present, there is no need to grow additional crops for alcohol. Present feedstocks to produce 660 million gallons of ethanol per year can come from cheese whey, citrus waste, corn, and grain sorghum. This amount could be pushed to about eight times, or 4.7 billion gallons per year, with existing biomass feedstocks, if production capacities were expanded, and MSW (municipal solid wastes) were brought into the feedstock stream.

Alcohol Production Costs

Ethanol. The price of ethanol, around $1.20 per gallon (1979), can be reduced to under $1.00 per gallon in large plants, if certain economies are effected:

- Continuous fermentation.
- Efficient distillation (vacuum) techniques.
- Economies of scale.
- Use of waste materials.
- Improved feedstock collection and by-product uses.

Methanol. In the 1980s, with coal as the feedstock, methanol should come down to 30 to 60 cents per gallon. This depends on heavy capital investments in facilities that are now nonexistent. Methanol can be more competitive with gasoline because of its higher efficiency in engines designed for its use.

Small-Scale Operations. Although the effects of large scale will reduce costs in general, ethanol can be produced economically by local farmers' organizations, when the costs of collecting, transporting, and storing feedstocks are kept down. Where sales are also limited to local areas, the product may go into Gasohol or as straight fuel for special farm equipment.

Net Energy Balance

The question of energy used in ethanol production was the first defensive tactic of petroleum-oriented criticism of the Gasohol movement. Older distilleries, using oil to generate process heat for production, with little or no recovery of heat inputs, were the butt of this complaint. However, there are many industries in which analysis of the heat balance is far less efficient than in old distilleries. Electric generation results in a loss of about 66 percent of input energy, usually to an ocean or river. In a modern plant, ethanol production can be positive in energy balance, and it can be done at low temperatures, with fuels other than oil or gas.

When alcohol fuel production uses minimal oil or natural gas, net energy balance need not be a concern.

Oil Replacement Value

The amount of oil that alcohol fuels displace equals the amount of oil they save or replace in particular uses, minus the amount of oil used to produce the alcohol. Used in Gasohol, ethanol (ethyl alcohol) saves oil in two ways:

As an octane enhancer and as a fuel. A refinery producing low-octane unleaded fuel (for alcohol-blending) can use less fuel in reforming than those producing higher-octane gasolines. If alcohol production reached 40,000 barrels per day by 1985, its octane-enhancing value alone would save about 2,400 barrels of oil per day.

As a fuel, ethanol contains roughly two-thirds the energy in gasoline.

The evidence on mileage improvement with Gasohol is varied and mixed. However, dynamometer tests by the DOE, the EPA, and university laboratories, indicate that Gasohol gives increased miles per Btu but decreased miles per gallon. Concentrations both lower and higher than 10 percent alcohol have merits.

One can also use pure alcohol, in engines redesigned with high compression ratios to utilize the high octane rating of alcohol. *Doing so yields greater miles per Btu than Gasohol. Cars optimized for pure methanol can attain 30 percent more miles per Btu than cars designed for gasoline or Gasohol.*

Automobile manufacturers say that such cars can be produced soon in small quantities for local fleet use, later in larger quantities, given the availability of fuel and the demand.

Environmental Impacts

Use of alcohol fuels can help improve the environment by recovering waste materials, such as paper pulps and cheese whey. Methanol from coal entails mining and transportation effects, known to be controllable. The first phase of methanol from solid fuel is gasification, with few effluents, and for which control techniques are known.

Overall, methanol production appears to be more environmentally benign than the combustion of coal to produce electricity.

The use of Gasohol generally decreases hydrocarbon and carbon monoxide emissions, slightly increases aldehyde and evaporation emissions. The EPA has permitted Gasohol use, based on the small amount marketed, and on the control of the emissions, in the early 1980s, by three-way catalysts, and larger cannisters for evaporative control.

This concludes the review of the physical properties of alcohol fuels and the inventory of applications up to the middle of 1979. The balance of the report, 85 percent of the whole, is devoted to policy initiatives, tax exemption incentives, tax credits, and four informative appendices. Anyone contemplating serious involvement with alcohol fuels will find the full report useful.

Experience with a Charcoal-Burning Gasogen

As an experiment, I designed and constructed a gasogen for a 1965 Chevrolet Malibu station wagon. The body style provided a level floor on which to place components and an "engineer's seat," looking backwards, for observation of instruments and manipulations by someone other than the driver. The station wagon also allowed the construction of an airtight safety-glass partition between the gasogen and the interior space. The rear window was fastened open.

The author's 1965 station wagon, equipped with an experimental gasogen. The drum is the main body of the gas generator.

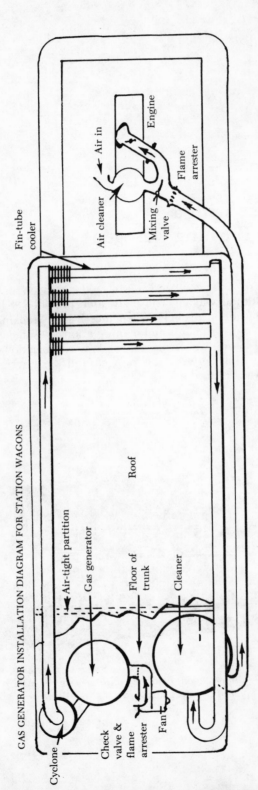

GAS GENERATOR INSTALLATION DIAGRAM FOR STATION WAGONS

Figure C-1. A diagram of the author's gasogen from above, showing the flow of gas from the generator to the cyclone and on to the air cooler. From the cooler, the gas travels back to the cleaner and then forward again to the engine.

Another view of the author's gasogen. The round object at lower left is the cyclone; at center are the check valve and flame arrester (not visible) and fan; and at right are flexible metal pipes to and from cleaner.

Experiences with Gasogens

The gasogen was modeled on the "early and heavy" style for charcoal or coal (see p. 48). These fuels are chemically active and form a porous fire bed, so that a single large nozzle suffices for air supply. The shell of the gasogen was a 15-gallon steel drum, used for shipping Freon. This choice limited the diameter of the *refractory liner* to 14 inches, so that the fire bed was inadequate for a big V-8 engine.

Some Initial Discoveries

My suspicions about the inappropriateness of gasogens for highway use were confirmed. The car's automatic transmission shared this feeling, and remained in low gear, unless going downhill. Some sophisticated remedies, involving engine and transmission changes, were rejected as too costly.

One positive discovery was that baseboard heating pipes, with fins attached, provided an adequate, low-cost cooling unit. We learned more about the combustion of damp wood chips than appears in books: they don't burn, unless mixed with about 75 percent charcoal briquets. We were shocked by the rapid corrosion of galvanized metal parts exposed to wood smoke at 900° F. The zinc melts away, if not eaten by the acid in the gases, and the steel is soon eaten away, too.

Solutions

The solution to these problems is within reach, but at the price of sub-stituting stainless steel for ordinary steel. Damp wood chips should be partly dried before charging the retort, and a special air supply must be used in a gasogen for wood fuel. Making joints between the components and the piping that are both airtight and easy to disassemble for cleaning presented a challenge. Asbestos gaskets will resist the heat, but require considerable compression to avoid leaking. Silicone-resin gaskets can be soft and tight, but may not last long in the hottest locations.

With more time, more experience, and an elastic budget, it would not be difficult to machine tight joints, either flanged or conical, to solve this

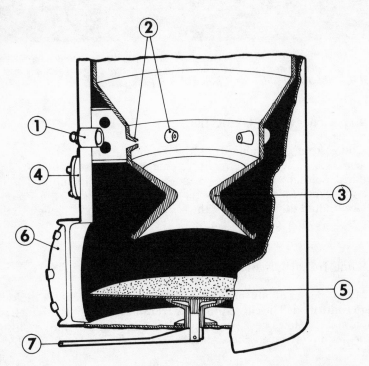

Figure C-2. A section of a Svedlund wood gasogen with these parts: (1) main air inlet and flame arrester (a screen inside cylinder), (2) nozzles, (3) hearth casting, (4) inspection and lighting port, (5) ash plate, (6) ash removal door, and (7) shaker handle.

worst problem of air leaks. Since the system is all under vacuum, leaks cannot be detected except by closing the carburetor by-pass valve (causing the engine to turn off), filling the whole system with smoke with the pressure blower, and watching for little plumes of smoke at joints. After marking the leaks, remember that the system now contains enough CO to blow the gasogen, the car, you, and perhaps near neighbors, to a point somewhat remote. It could also poison a number of visitors if confined in an unventilated space. The way to make this operation of leak detection safe, as well as preventing pop-backs or worse explosions, is to insert fine-mesh copper screens in the prescribed locations in the piping. One is shown in the diagram of the Svedlund gasogen (Figure C-2), and called a "flame arrester."

Consider This

Two comments, from different parts of the world, are passed on to prospective researchers: A Japanese firm dealing in small technological equipment, including gasogens, urged their customers for small, four-to-ten horsepower engine sets to buy two gasogens, so that cleaning one unit would not interrupt operation of the engine. The Swedes, who so successfully kept transportation active with gasogens during World War II, are now embarked on a program to produce methanol fuel from their forest resources, at far greater efficiency than by gasogens.

BUILD A STATIONARY UNIT

There are certain advantages to building one's first gasogen as a stationary unit. For one, weight is of no importance, so use steel of no less than 16 gage. It lasts longer than the light shipping drums often used by amateur builders. The firebox for gasogens that will use charcoal or coal can be made of a refractory. This will outlast sheet metal and save the expense of double-walled vessels. However, a mold for the refractory is required. I used Johns-Manville's Lightweight Firecrete, available from commercial insulators and boiler installers, in bags of 85 pounds. It is mixed with water and looks like ordinary cement, but resists temperatures up to 2400° F. Another brand, "Narcocast" is also suitable, but heavier.

Grate

A cast iron grate must be mounted so as to allow ashes and clinkers to be dumped. I found one from a stove in a junk yard, repaired a broken segment with epoxy-steel mixture, and had a foundry cast a new one, using the old as a pattern. The ring that supports the grate was flame cut from ⅛" steel plate, as were the stirrups that hold the rocker lugs on the grate. A pipe handle slips over one of these, through the ash door, to shake the grate or to dump it.

Nozzle

The nozzle is a forged reducer fitting, stocked by piping dealers. Welding the heavy nozzle pipe fitting to the relatively thin sheet metal of the retort and the cleaning and cooling vessels is difficult. So use pads or flanges where the pipe contacts the sheet metal, with bolts or studs clamping the parts together. Sleeves with asbestos packing join tubing to the vessels. The sleeves can be made of light-gage metal.

Cyclone

The "cyclone" is a device used in many industries to separate solids from a gas moving circumferentially inside it. Particles of fuel and ashes, being heavier than our producer gas, are flung out by centrifugal force and fall to the bottom. This is a pan held in place by four luggage latches, which exert some force on the gasket. The clean-out design on the Japanese gasogen (Figure C-3, part E) is superior, and should be followed. The small threaded cup makes a much tighter joint than the large diameter gasket.

Gas Cooler

The next component in the system is the gas cooler. In an automotive unit, the required cooling surface area reduces streamlining and car stability. On my station wagon, the rooftop seemed to be the best location. However, at slow speeds, there was inadequate air turbulence over the finned tubing, and gas temperatures remained high. Remember, an automobile's radiator passes heat from a liquid to a gas (air), but the gasogen's cooler is a gas-to-gas exchanger, in which efficiency is very poor.

In a stationary gas plant, the cooler may be a drum in which water is sprayed directly into the gas stream. To minimize the use of water, a circulating pump and cooling basin must be considered. After a day's use, the water will be laden with dissolved and particulate impurities, and will have to be dumped where there is no chance of contaminating water supply or agricultural land.

The air cooler in the Japanese wood gasogen system (Figure C-3), while not efficient in terms of size, is easy to make. It depends, for part of its cooling, on natural convection of air rising through the central flue. Hence its location in an open space is important.

At left, a close-up of a baseboard unit used for the gas cooler; at right, the cyclone cleaner secured with luggage latches.

This cast refractory firebox for charcoal-fired gasogen (left) was abandoned because of its excessive weight (80 pounds). At right are baseboard units and header pipe for gas cooler.

133

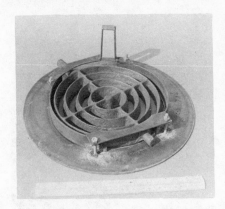
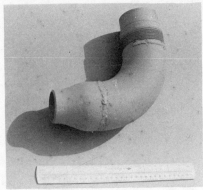

Top left: grate, upside down, showing support straps that permit shaking and dumping ashes. Top right: nozzle for charcoal gasogen. Left: grate in ash-dumping position.

Here, author John Lincoln checks the gasogen in his 1976 Chevrolet.

Filter

The filter, or scrubber, is important to the life of the engine, which will operate quite well without it. Because frequent changes of the filter medium are necessary, the hardware must be simple and reliable. On the cannister in our ideal unit, the cover is removed by loosening a half-dozen wing nuts, or wing bolts, around the rim. The gas at this point should be cooled to about 30 to 40° C. (86 to 104° F.) and rubber or silicone gasket material will be adequate. Various filter materials may be used, depending on the need for cost, expendability, and absorptive quality.

Many Swedish gasogens used cloth filters, which were most effective in removing the fine particles from charcoal and wood gasogens. They were washable and durable at the correct temperatures and, when properly designed in accordian pleats, they were the most compact and lightest filters. Industrial air conditioning supply houses may carry the materials in stock.

On my Chevrolet conversion, the air cleaner consisted of three trays of expanded metal lath. In succession, gas went through shredded bark, charcoal, and fiberglass insulation. The charcoal was recycled, but the others were dumped every three to five firings, regardless of mileage. The tubing from the cleaner lid to the mixing valve may be flexible metal hose or composite rubber and wire, both of which will resist collapse under vacuum.

A WOOD-BURNING GASOGEN

The sizes of the components of four small wood-fired gasogen systems are given in the tables accompanying Figure C-3. The pipe sizes are applicable to wood systems, but gasogen dimensions for charcoal fuel may be scaled down somewhat, due to the higher calorific density of charcoal. Coolers and filters may be sized to conform to ready-made tanks, barrels, or tubes that the builder may have or acquire on the market.

Note the ample radius shown on each pipe bend. It is very important, in low-pressure piping, to avoid sharp bends or restrictions of any kind. Use the maximum radius possible, and none less than twice the diameter of the pipe.

Cut the Right Wood

The gasogen builder's first decision must be what fuel to use. Charcoal gasogens are simple to construct, with a single nozzle, but a supply of charcoal, with no more than 10 to 12 percent moisture, may be difficult to purchase at a reasonable price. Such a supply was assured in Sweden by

SMALL STATIONERY GASOGEN. DIMENSIONS, MILLIMETERS AND INCHES

Size	N	O	P	Q	R	S	T	Engine HP
1	350	900	150	350	200	40	150	18–20
	13¾	35½	6	13¾	7¾	1½	6	
2	300	900	150	300	200	40	150	12–15
	11¾	35½	6	11¾	7¾	1½	6	
3	300	700	120	300	180	32	100	10–12
	11¾	27½	4¾	11¾	7	1¼	4	
4	250	650	120	250	180	32	100	4– 8
	9⅞	25½	4¾	9⅞	7	1¼	4	

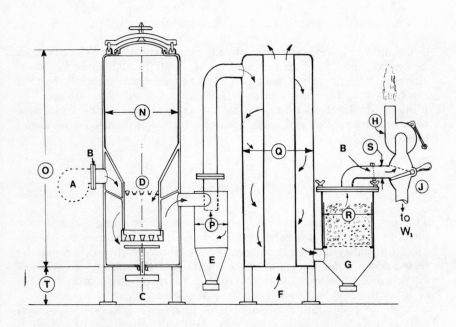

KEY

A Pressure blower (alternate)
B Flame arrester
C Gasogen
D Tuyeres
E Cyclone
F Cooler
G Purifier
H Suction blower (preferred)
J Starting valve

Figure C-3. In this gasogen system for wood, ease of fabrication and light weight, not efficiency, dictated design. It was for use in mountains of Japan, for sawmills, etc. Wood chips or cocoa shells of maximum dimension of 30-mm. (1¼") were used. Data from "Cecoco," Osaka, Japan.

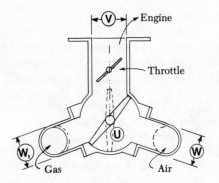

MIXING VALVE DIMENSIONS

Type	U	V	W	
A*	70	40	40	(mm.)
	2¾	1⁹⁄₁₆	1⁹⁄₁₆	(in.)

*Used with gasogens size 1–4

Figure C-4. Detail of a gas-air blending valve with dimensions above.

government action. The more complex wood gasogen's fuel can be cut with ordinary power saws. Maximum moisture content is from 20 to 30 percent, and size, to feed by gravity without constant attention, is about one inch square by two to four inches long. The size suggested for the units shown is about 30 millimeter cubes.

Getting Started

A description of starting a wood gasogen for the first time, or after a cleaning and overhaul session, gives some indication of the expertise gained in Sweden during the war. Over the grate place coke or dense charcoal, and porous concrete, in equal parts. (For the concrete, substitute broken bricks of refractory from a discarded oil burner. The lumps should be just large enough to avoid lodging in the grate.) These lumps are reusable, acting as a heat sink and diffuser for the formation of carbon monoxide (CO). Fill up to within about four inches of the nozzle ring. Then fill with light charcoal to about three to five inches above the nozzles. On top of this, fill the gasogen about ⅓ full of dry wood (10 to 12 percent moisture) and fill the balance with as-cut wood.

Lighting is usually by a gas torch (Liquid Petroleum gas) through the hole provided near the throat, or nozzle ring. The suction fan, which is preferred to the hand blower (Figure C-3), is turned on simultaneously, venting to the atmosphere. In about six to twelve minutes, with a torch again, light the stream of gas at the bypass valve. The flame should be a steady red-blue jet. White or gray mist at the core indicates that there is too much moisture in the fuel. Stop the blower, close the testing valve, and start the engine (the gasoline carburetor is removed, in this case) in the usual way. The gas-air mixing valve is adjusted for the best speed with a fixed low idling setting of the throttle valve.

The Process

It is advisable to understand what happens in the gasogen. From the top down, the wood is dried, then charred, releasing water, carbon dioxide (CO_2), CO, tars, acetic acid, etc. The distillation gases are burned, with the charcoal; finally the tars are decomposed (presumably to carbon and hydrogen) and the CO_2 is reduced to CO. It is apparent that by burning charcoal, the process is made simpler and more efficient.

The engine may be stopped for 10 or 15 minutes without losing the fire. But note that during the stopped period the generator continues to produce CO, and will no longer be under vacuum. Take care not to allow gas to escape at this stage. It is odorless, lethal, and very explosive. When the engine is needed, try a few revolutions of the starter. This failing, run the suction blower a few minutes, testing the gas with a torch as before.

If ignition fails to occur, the fire is out or beyond recovery, and the system must be purged of gas before relighting. Note that the flame arresting screens protect both the inlet and outlet ends of the system from flames outside the system. These may be backfires from the engine or sparks. However, the lighting torch flame, introduced into the gasogen, full of air and CO, could have tragic results. Open the fuel filler lid and run the blower until you are sure that air alone fills the system. Then close the lid, and start from scratch with the torch.

Legal Problems

There are several cautions to observe when using gasogens in automobiles on highways. In addition to the inconveniences mentioned, and the loss of power and acceleration, there is a legal problem to which we can find no easy technical solution. The pollution control devices on American cars and most of the imports depend on taking in excess air. To provide a combustible gas from a gasogen, these air intakes, including the crankcase blow-by connection, must be plugged. But this is illegal according to federal environmental regulations. The first result of plugging any pollution control devices will be to detune the carburetor for running on gasoline. With my old Chevrolet, the next result was a stream of blue smoke from my exhaust, an irresistible signal for a state police cruiser to appear in my rear view mirror.

Suggested Improvements

Although the Swedes developed effective changeover hardware for dual-fuel vehicles, the complications of many cables and interlocking control rods will not appear to be warranted until the price of gasoline reaches many times its 1980 level. The experimenter will find more scope for improvement of engines destined to run solely on gas generated on board. These improvements, suggested by Swedish writers, are as follows:

1. Increase compression ratio by "stroking," or shaving the head, milling pistons, etc.

2. Supercharging. Turbo-superchargers are available for many sports cars now. These restore power that the diluted gasogen fuel lacks by pumping more of it into the cylinders each stroke.

3. Lowering the final drive ratio. This may be the least expensive solution. Stock differential gears in several ratios are usually available for passenger cars and trucks. However, wear of the engine and transmission will be increased and maximum speeds will be restricted.

Experimenters may find a wealth of information in a translation of the standard Swedish book on the subject.

Generator Gas: The Swedish Experience from 1939–1945. Translated by the Solar Research Energy Institute. Book No. SERI/SP-33-140. January, 1979

Available from:
National Technical Information Service
U.S. Department of Commerce
5285 Port Royal Road
Springfield, VA 22161
Microfiche $ 3.00   Printed copy $12.00

A Survey of Biomass Gasification. Vol. 2. Solar Energy Research Institute (SERI).

Available from:
National Technical Information Service
U.S. Department of Commerce
5285 Port Royal Road
Springfield, VA 22161
Microfiche $3.00   Printed copy $9.50

Volume 1 of the above is a summary.
Microfiche $3.00   Printed copy $4.50.

COMPONENTS USEFUL IN BUILDING WOOD AND
CHARCOAL-BURNING GASOGENS

Flexible metallic tubing. It is very difficult to bend pipe over an inch in diameter around compound curves, regardless of material. A corrugated copper tubing used in solar heating plumbing will withstand moderate temperatures and the 20-inch vacuum expected in gasogens. Called "Hydro-Flex," available from plumbing supply houses, and Hydro-Flex Corp., 2101 N.W. Brickyard Rd., Topeka, KS 66618.

Hand-driven blowers. While electric blowers for operation on 12 volts D.C. are available from many industrial and marine supply houses, the only hand-cranked one I could find was made by Buffalo Forge Company, P.O. Box 895, Buffalo, NY 14240.

Complete Gas Generating Systems, Designed and Made on Order

ECON (The Energy Conservation Co.)
Alexander City, AL 35010
Ben Russell, President

Developing a modular unit for trucks and stationary uses.

Fuse-Weld, Inc.
1052 East 43 St.
Hialeah, FL 33013

Harry La Fontaine, consultant

Imbert Air Gasifier,
Steinweg Nr. 11
5760 Arnsberg 2, Germany

Down-draft gasifiers for diesel-electric generator sets, 10 kw to 10,000 kw sled or trailer.

Vermont Wood Energy Corp.
P. O. Box 280
Stowe, VT 05672
Peter H. Bauer, project engineer

Developing residential furnace-size gasifier for wood chips or pellets.

Listing does not imply endorsement. Many firms and individuals have shown great interest in portable gasogens. Usually they build one prototype, and then withdraw from further activity.

Glossary

AAAS: American Association for the Advancement of Science.

API: American Petroleum Institute.

BTU: British thermal unit. Heat required to raise 1 lb. of water 1 degree Fahrenheit. MBtu = 1,000 Btu. 1 = 0.25 calorie.

C: Symbol for carbon.

CAL: Calorie. Heat required to raise 1 gram of water 1 degree centigrade.

CI: Compression ignition (engine). A diesel.

CO: Carbon monoxide, a combustible toxic gas.

CO_2: Carbon dioxide, an incombustible inert gas, found in the exhalation of all animals.

CVCC: Controlled Vortex Combustion Chamber.

DIESELING: The English call it "running on." A gasoline engine, with the ignition shut off, will sometimes keep running. Some fuel gets into the cylinder through the idling valve, and the oil on the cylinder wall is fuel, which ignites due to the heat added by compression of air. Timing is random, and undue strains result. Alcohol in the fuel usually absorbs the heat, prevents the ignition, and the running.

DISSOCIATION: The process by which a chemical combination breaks up into simpler constituents, usually capable of recombining under other conditions. For example, methanol may dissociate thus: $CH_3OH \longrightarrow 2 H_2 + CO$, with the absorption of heat.

DOE: Department of Energy.

DOT: Department of Transportation.

EPA: Environmental Protection Agency.

ETHANOL: A grain alcohol, also known as ethyl alcohol or C_3H_5 OH.

141

FEEDSTOCK: Raw material for alcohol production. Feedstocks for ethanol are grains, sugar canes, tubers, tapioca, garbage. Feedstocks for methanol are natural or coal gas, wood, corn cobs, straw, anything rich in carbon.

GASOGEN: Short form for gas generator, the portable or stationary small units that enabled many European drivers to convert their vehicles to a solid fuel.

GASOHOL: A blend of 90 percent lead-free gasoline with 10 percent ethyl alcohol (by volume). The latter is usually specified to be 200 proof, or anhydrous, without water.

GASOLINE: A volatile liquid hydrocarbon mixture, often containing a dozen compounds. May be natural, straight-run, cracked, or a combination.

HC: Hydrocarbon. Gases or liquids containing hydrogen and carbon.

IC: Internal Combustion.

IPAA: Independent Petroleum Association of America.

H: Hydrogen.

H_2O: Water.

HP: Horsepower. (1 Hp = 2545 Btu per hour.)

KCAL.: Kilogram-calorie, equal to 3.9 Btu (calories are more fattening).

KNOCK: A combustion phenomenon. The spontaneous combustion, usually premature, of a major part of the charge, due to many causes. It means that power is being lost.

LPG: Liquified petroleum gas.

METHANOL: A wood alcohol, made from coal, wood waste or any material containing carbon; known by the formula CH_3OH.

MON: Motor octane number.

NO_X: Nitrogen oxides, a pollutant of most IC engines.

O: Oxygen.

OPEC: Organization of Petroleum Exporting Countries.

RON: Research octane number.

SYNFUEL: A recently coined word, intended to mean synthetic fuel. The trouble is, nobody can agree on what definition to use. It will be more useful to categorize fuels as to origin: from fossil sources, or from renewable sources. Synthetic gasoline, for example, might be made from coal or oil shale; it might be methanol, made from wood wastes (renewable crop) or from ethanol (from grains or fermentable wastes of other renewable crops).

Bibliography

Alcohol-Gasoline Studies at Massachusetts Institute of Technology. No. 3, API series, 1941.

Coates, Joseph F. (Office of Technology Assessment, U.S. Congress). "Long-Range Trends and the Future of the Automobile in America." Paper presented to symposium on the Future of Cars, AAAS, Annual Meeting, Jan. 28, 1975.

Davis, William M. *Designing a Steam Car.* Stonington, Conn.: New Steam Age Publishing, 1942.

Derr, Thomas S. *The Modern Steam Car and Its Background.* Newton, Mass.: First edition privately published by American Steam Automobile Co., 1932. Reprinted and supplemented by Clymer Books, Los Angeles, California.

Drew, Elizabeth. (A Reporter at Large). "The Energy Bazaar." *The New Yorker*, July 21, 1975.

Egloff, Dr. Gustav. *Motor Fuel Economy of Europe.* API Series, No. 2, 1939.

"EV (Electric Vehicle) Revival: A Special Report." *Machine Design*, October 17, 1974.

Farmer, Weston. "Old Fuel System May Be Answer to Costly Oil." *National Fisherman*, June–July, 1975.

Halacy, Daniel S., Jr. *Fuel Cells: Power for Tomorrow.* Cleveland, Ohio: World Publishing Co., 1966.

Hale, Lucretia P. *The Peterkin Papers.* Boston: Houghton Mifflin Co., 1886.

Hall, J. Alfred. "Wood, Pulp and Paper, and People in the Northwest." Joint report of Northwest Pulp and Paper Association, Seattle, Wash., and Pacific Northwest Forest and Range Experiment Station Forest Service, Portland Oregon: U.S. Department of Agriculture.

Hammond, Allen L., Metz, William D., and Maugh, Thomas H., II. "Energy and the Future." American Association for the Advancement of Science, 1973.

Harvey, Douglas G. and Menchen, W. Robert. *The Automobile — Energy and the Environment. A Technology Assessment of Advanced Automobile Propulsion Systems.* (NSF Contract, RANN, No. NSF-C674) Columbia, Maryland: Hittman Associates, Inc., March, 1974.

Heitland, Herbert; Bernhardt, Winfried and Lee, Wenpo. "Comparative Results on Methanol and Gasolene Fueled Passenger Cars." R & D Division, Research Department, Volkswagenwerk AG, Wolfsburg, Germany, 1974.

Homans, James E., A.M. *Self-Propelled Vehicles. A Practical Treatise on the Theory, Construction, Operation, Care and Management of All Forms of Automobiles.* New York: Theo. Audel and Company, 1911. Esp. denatured alcohol, pp. 252, 253, 254, Fig. 180.

Hornby, John. *A Text Book of Gas Manufacture.* London: 1900.

Instruction Book for Government Utility Gas Producer, Marks VI and VII. London: HMSO, Ministry of War Transport, 1943.

Lerner, R.M. et al. "Improved Performance of Internal Combustion Engines Using 5–20 Percent Methanol." Lincoln Laboratory, MIT, (n.d.)

Ludvigsen, Karl. "Automobile Aerodynamics, Form and Fashion." *Automobile Quarterly*, vol. VI, no. 2 (1967).

Mancke, Richard B. *The Failure of U.S. Energy Policy.* New York: Columbia University Press, 1974.

Medvin, Norman. *The Energy Cartel: Who Runs the American Oil Industry?* New York: Vintage Books, Random House, 1974.

"Methanol as an Alternate Fuel." Reprints of 1974 Engineering Federation Conference, New England College, Henniker, N.H., July, 1974. (2 volumes)

Miller, Conrad. "Beating the Energy Pinch." *Motor Boating & Sailing*, March, 1974.

Myles, Bruce. "Now You Can Read Your Newspaper and Eat It, Too. Chemical Transforms Cellulose into Food and Fuel." *Christian Science Monitor*, March 4, 1974.

Nash, Alfred W. and Howes, Donald A. *The Principles of Motor Fuel Preparation and Application.* New York: John Wiley & Sons, Inc., 1935. Vol. I, "Alcohol Fuels," pp. 349–351; vol. II, "Motor Fuel Specifications."

On the Trail of New Fuels: Alternative Fuels for Motor Vehicles. Translated from German. Lawrence Livermore Laboratories, University of California, Livermore, Cal., June, 1975.

"The Prospects for Gasoline Availability: 1974." A background paper prepared by the Congressional Research Service at the request of Henry M. Jackson, chairman, Committee on Interior and Insular Affairs, U.S. Senate. Serial No. 93–41 (92–76) Washington, D.C.: U.S. Government Printing Office, 1974.

Reed, Thomas B. "Biomass Energy Refineries for Production of Fuel and Fertilizer." Cambridge: Massachusetts Institute of Technology. Paper presented at Eighth Cellulose Conference, May, 1975.

Reed, Thomas B. and Lerner, R.M. "Methanol: A Versatile Fuel for Immediate Use." *Science*, Dec. 28, 1973. vol. 182, pp. 1299–1304.

Reed, Thomas B. and Lerner, R.M. "Sources and Methods for Methanol Production." Lincoln Laboratory, MIT (n.d.)

Roberson, Edwin Cecil and Herbert, Roy. *Fuel: The Conquest of Man's Environment.* New York: Harper & Row, 1963. 128 pp.

Rouyer, G. *Etude des Gazogenes Portatifs.* Paris: Dunod, 1938.

Scott, Robert F. "A Sleeping Giant Awakes: The Invention of the Internal Combustion Engine." *Automobile Quarterly*, vol. VI, no. 4, Spring, 1968. p. 410.

Technical Characteristics of Alcohol-Gasoline Blends. American Petroleum Institute. Motor Fuel Facts Series, no. 1, 1938.

Trotter, Robert J. "Is Hydrogen the Fuel of the Future?" *Science News*, vol. 102, July 15, 1972.

U.S. Department of Transportation. *Cost of Operating an Automobile.* Federal Highway Administration, Highway Statistics Division. Washington, D.C.: Government Printing Office, April, 1972.

U.S. Department of Transportation and the U.S. Environmental Protection Agency. "Potential for Motor Vehicle Fuel Economy Improvement." Report to the Congress, 24 October 1974.

Wigg, E.E. (senior research chemist, Exxon Research and Engineering Co., Linden, N.J.). *Science*, Nov. 29, 1974. vol. 186, no. 4166.

Willkie, Herman Frederick and Kolachov, Dr. Paul John. *Food for Thought.* Indianapolis, Ind.: Indiana Farm Bureau, Inc., c. 1942. (PPL No. B662.6 W 73f.)

Index